PhoneGap
Beginner's Guide

Build cross-platform mobile applications with the PhoneGap open source development framework

Andrew Lunny

PACKT PUBLISHING

BIRMINGHAM - MUMBAI

PhoneGap

Beginner's Guide

First published: September 2011

Production Reference: 1160911

Published by Packt Publishing Ltd.
Livery Place
35 Livery Street
Birmingham B3 2PB, UK.

ISBN 978-1-849515-36-8

www.packtpub.com

Cover Image by Asher Wishkerman (a.wishkerman@mpic.de)

Credits

Author
Andrew Lunny

Reviewers
Paul McCrodden
Andrey Rebrov

Acquisition Editor
Usha Iyer

Development Editor
Meeta Rajani

Technical Editors
Pallavi Kachare
Priyanka Shah

Project Coordinator
Joel Goveya

Proofreader
Aaron Nash

Indexer
Hemangini Bari

Production Coordinator
Aparna Bhagat

Cover Work
Aparna Bhagat

About the Author

Andrew Lunny is a software developer based in Vancouver, BC, where he is "Chief N00b" at Nitobi Software. He has worked at Nitobi for four years, since a brutal shark attack cut short his promising career as a surfer. He is the lead developer and all-around fall guy for the PhoneGap Build web service, a member of the PhoneGap team, and has over 10 years' experience with PhoneGap and related technologies. He is fond of Ruby, JavaScript, Unix, Git, and the Internet.

Nitobi is a software company run by Andre Charland, Dave Johnson, and Brian Leroux. They specialize in cross-platform mobile development and design, and sponsor the PhoneGap open source project.

In his spare time, Andrew enjoys cycling, running, walking, and jumping. He has two unrelated degrees from the University of British Columbia.

Thanks to Tammy, to my parents, and to Hugo and Natasha for putting up with me while I procrastinated along. Thanks to Michael Brooks for running a Windows VM so I did not have to, and thanks to all of my other co-workers for working on such a great open source project. Thank you to everyone who follows me on Twitter.

Finally, thanks to all my old surfing buddies—Skip, Dingo, even Jelly! I'll get back on the waves with you guys some time, I promise.

About the Reviewers

Paul McCrodden is a Digital Media Developer from a small country in the west of Europe known for its potatoes and leprechauns. He is a chartered engineer graduating with a Bachelor of Engineering in Digital Media from Dublin City University and a Master of Science in Multimedia Systems from Trinity College Dublin. With this knowledge he aims to merge the technical with the creative when it comes to digital media production.

Paul has previously worked for global companies such as Ericsson and Bearingpoint Consulting along with various medium to small businesses on contract. He recently gave up the 9 - 5 to work 9 - 9 for his own company, which provides digital media consulting and a range of services including, web solutions, mobile application development, and other multimedia. For more information go to `www.paulmccrodden.com`.

I would like to thank my family, friends, and last but by no means least, my girlfriend Claire for her support and patience. A huge thanks to the Drupal community for all their hard work and the Wordpress and PhoneGap communities for theirs. Lastly, a thank you to Packt for asking me to review the book; it has been a great all round learning experience.

Andrey Rebrov, 23, has big plans for his future. He started as a software developer in Magenta Technology—a big British software company specialized in Enterprise Java Solutions, and worked there for more than three years. Now, he is working on Luxoft as a senior Java web developer. In February 2011 he graduated from the IT Department of Samara State Aerospace University with an honors degree.

He is also working with different web technologies—JavaScript (ExtJS, jQuery), HTML/HTML5, CSS/CSS3, and has some experience in Adobe Flex. He is also interested in mobile platforms, and developing web and native apps. Exploring PhoneGap, jQuery Mobile, and Sencha Touch, he has created some plugins for jQuery Mobile.

In his work he uses Agile methodologies such as Scrum and Kanban, and is interested in innovation and Agile games. At the moment he is working on innovation games popularization in Russia and Russian communities.

He also uses GTD methodologies, which help him in his work and daily routine. And, of course, it helps him with blogging.

In his blog he writes about interesting technologies, shares his developer experience, and translates some articles from English into Russian.

I would like to thank Kartikey Pandey who asked me to review this book and Joel Goveya who helped me throughout the review.

I would like to thank my parents for providing me with the opportunity to be where I am. Without them, none of this would even be possible. You have always been my biggest support and I appreciate that.

And the last thanks go to my girlfriend, Tatyana, who always gives me strength and hope.

Table of Contents

Preface

PhoneGap: A Beginner's Guide is an introduction to PhoneGap: an open source, cross-platform framework for developing mobile applications. PhoneGap allows developers to leverage web development skills—HTML, CSS, and JavaScript—to developed native applications for iOS, Android, BlackBerry, and many other platforms with a single codebase. Many of the same benefits of developing websites—for example, deployment to a wide variety of clients—are at developers' fingertips.

What this book covers

Chapter 1, Installing PhoneGap, helps readers through the often difficult process of setting up multiple development environments for the iOS, Android, and BlackBerry platforms. After this chapter, you will have an environment ready to build your PhoneGap applications.

Chapter 2, Building and Debugging on Multiple Platforms, shows how to use the environment set up in *Chapter 1* to quickly and efficiently work on your code for multiple platforms at once. It also helps you get used to using desktop browsers to assist with mobile development.

Chapter 3, Mobile Web to Mobile Applications, describes the changes in application design and architecture that are at the forefront of developing on PhoneGap. In particular, we see how to write PhoneGap applications that do not rely on a web server for the majority of their interactions.

Chapter 4, Managing a Cross-Platform Codebase, shows readers how to use common web techniques, including feature detection and user-agent sniffing, to manage their code that gets deployed to multiple platforms.

Chapter 5, HTML5 APIs and Mobile JavaScript looks at some of the new JavaScript APIs available in HTML5 browsers, which are common on modern mobile devices. We also look at mobile JavaScript libraries that are useful for managing your code.

Chapter 6, CSS3: Transitions, Transforms and Animation, looks at the new techniques available in current CSS implementations for sprucing up the look and feel of your PhoneGap applications.

Chapter 7, Accessing Device Sensors with PhoneGap, demonstrates the use of PhoneGap's device sensor capabilities for managing the location and accelerometer readings from your PhoneGap application.

Chapter 8, Accessing Camera Data and Files, shows how to use the PhoneGap APIs to manage access to the user's photo library and camera, and use the results in your application.

Chapter 9, Reading and Writing to Contacts, uses the Contacts APIs from PhoneGap to work with the user's native contacts list on their device, for use in your own application.

Chapter 10, PhoneGap Plugins, shows how the iOS, Android, and BlackBerry implementations of PhoneGap can be easily extended to access any native capabilities not exposed by the PhoneGap core APIs.

Chapter 11, Working Offline: Sync and Caching, shows how with a small amount of server-side code, you can use PhoneGap to capture data offline and manage it locally or remotely.

Appendix A, Deploying to iOS, shows you how to get a Developer Certificate from Apple, allowing you to take your application from a simulator to the market.

What you need for this book

Since PhoneGap uses the native capabilities of each supported mobile platform, you will need to install the appropriate native SDKs for each platform that you want to deploy to.

In the case of iOS, you will require an Apple Mac computer. For BlackBerry, you will require a Windows PC, or a virtualized Windows environment. Android's SDK supports all major operating systems.

Other than that, you will just need a web browser—preferably a WebKit based one, such as Safari or Google Chrome—and a text editor.

Who this book is for

This book is ideal for intermediate web developers, who have not worked on the mobile web or on mobile applications. No experience with native mobile SyDKs is required.

Conventions

In this book, you will find several headings appearing frequently.

To give clear instructions of how to complete a procedure or task, we use:

Time for action – heading

1. Action 1
2. Action 2
3. Action 3

Instructions often need some extra explanation so that they make sense, so they are followed with:

What just happened?

This heading explains the working of tasks or instructions that you have just completed.

You will also find some other learning aids in the book, including:

Pop quiz – heading

These are short multiple choice questions intended to help you test your own understanding.

Have a go hero – heading

These set practical challenges and give you ideas for experimenting with what you have learned.

You will also find a number of styles of text that distinguish between different kinds of information. Here are some examples of these styles, and an explanation of their meaning.

Code words in text are shown as follows: "From your `FirstApp` directory, launch the simulator."

A block of code is set as follows:

```
<meta name="viewport" content="width=device-width, initial-scale=1.0,
user-scalable=no"></meta>
```

Any command-line input or output is written as follows:

```
$ ant load-device
```

New terms and **important words** are shown in bold. Words that you see on the screen, in menus or dialog boxes for example, appear in the text like this: "To find the PhoneGap Sample, hit the BlackBerry button (the one with seven circles resembling a **B**), then **Downloads**".

> Warnings or important notes appear in a box like this.

> Tips and tricks appear like this.

Reader feedback

Feedback from our readers is always welcome. Let us know what you think about this book—what you liked or may have disliked. Reader feedback is important for us to develop titles that you really get the most out of.

To send us general feedback, simply send an e-mail to `feedback@packtpub.com`, and mention the book title via the subject of your message.

If there is a book that you need and would like to see us publish, please send us a note in the **SUGGEST A TITLE** form on `www.packtpub.com` or e-mail `suggest@packtpub.com`.

If there is a topic that you have expertise in and you are interested in either writing or contributing to a book, see our author guide on `www.packtpub.com/authors`.

Customer support

Now that you are the proud owner of a Packt book, we have a number of things to help you to get the most from your purchase.

Downloading the example code

You can download the example code files for all Packt books you have purchased from your account at `http://www.PacktPub.com`. If you purchased this book elsewhere, you can visit `http://www.PacktPub.com/support` and register to have the files e-mailed directly to you.

Errata

Although we have taken every care to ensure the accuracy of our content, mistakes do happen. If you find a mistake in one of our books—maybe a mistake in the text or the code—we would be grateful if you would report this to us. By doing so, you can save other readers from frustration and help us improve subsequent versions of this book. If you find any errata, please report them by visiting `http://www.packtpub.com/support`, selecting your book, clicking on the **errata submission form** link, and entering the details of your errata. Once your errata are verified, your submission will be accepted and the errata will be uploaded on our website, or added to any list of existing errata, under the Errata section of that title. Any existing errata can be viewed by selecting your title from `http://www.packtpub.com/support`.

Piracy

Piracy of copyright material on the Internet is an ongoing problem across all media. At Packt, we take the protection of our copyright and licenses very seriously. If you come across any illegal copies of our works, in any form, on the Internet, please provide us with the location address or website name immediately so that we can pursue a remedy.

Please contact us at `copyright@packtpub.com` with a link to the suspected pirated material.

We appreciate your help in protecting our authors, and our ability to bring you valuable content.

Questions

You can contact us at `questions@packtpub.com` if you are having a problem with any aspect of the book, and we will do our best to address it.

1
Installing PhoneGap

PhoneGap is, at heart, a set of project templates for different mobile operating systems, allowing us to ignore the details of each SDK and develop applications in a consistent fashion. The biggest roadblock developers find with PhoneGap is getting started—installing the SDKs and development environment to run those project templates. We're going to cover that task in this chapter.

In this chapter, we will:

- Install Xcode and the iOS SDK for iOS development
- Install the Android SDK and set up the emulator
- Set up the BlackBerry Web Works development environment
- Build PhoneGap applications to all three platforms

So let's get to the first roadblock...

Operating systems

We touched on this in the preface, but it's worth emphasizing again: PhoneGap plays by the rules. If a vendor releases their SDK for just a single operating system, then you will have to use that OS to build and deploy your applications.

In detail, for each PhoneGap platform:

- You can develop **Android** and **HP webOS** apps on any of the major desktop operating systems—Windows, Mac OS X, or Linux. Hooray!

- You can develop **Symbian Web Runtime** apps on any OS, but you can only run the simulator from Windows.
- Developing for **BlackBerry** is similar—the SDK can be installed on Windows or Mac OS X, but, at the time of writing, the simulator only runs on Windows.
- The **Windows Phone 7** SDK only runs on Ubuntu Linux, versions 10.04 and above. That one's a joke.
- And, as you're no doubt aware, the **iOS SDK** requires the latest version, 10.6, of Mac OS X (and, according to the OS X EULA, a Mac computer as well).

Practically speaking, your best bet for mobile development is to get a Mac and install Windows on a separate partition that you can boot into, or virtualize using **Parallels** or **VMWare Fusion**. According to Apple's legal terms, you cannot run Mac OS X on non-Apple hardware; if you stick with a Windows PC, you will be able to build for every platform except iOS.

If getting a new computer sounds a bit too expensive, you'll probably want to skip the *How To Buy a Dozen Phones* chapter as well.

Dependencies

Nearly there! There are just a few other important tools you'll need to get up and running:

- **git**: git is the best thing in the world (other opinions are available). More precisely, git is a distributed version control system, a superb tool for managing every aspect of software development. PhoneGap is developed with git at hand, and git is the best way to work with PhoneGap. You can find installation and usage information at `http://git-scm.com/`.

  For Mac and Linux users, I recommend using Git directly from the Terminal application. If you're on Windows, you may consider MsysGit, available at `http://code.google.com/p/msysgit/`.
- **ant**: More precisely Apache Ant, is a Java-based build tool, similar to make. Unlike make, ant tasks are specified with XML. It is a very popular tool in the Java community. We'll use ant extensively for building Android and BlackBerry apps. You can get ant on your system from `http://ant.apache.org/`, Detailed and current installation instructions are available there.

  Also, ensure that `ANT_HOME` is set correctly in your environment variables—this will ensure the PhoneGap ant scripts can run correctly.
- **Ruby**: The droidgap build tooling depends on Ruby, a widely available programming language. If you're running on Mac OS X or Linux, Ruby should be available with your installation. An installer for the latest Ruby release for your system is available at `http://www.ruby-lang.org/en/downloads/`.

Getting started with iOS

So let's finally get something started—building applications on iOS. Firstly, we're going to get the developer tools installed on our Mac, and then we'll get PhoneGap itself up and running.

Time for action – Getting an app running on the simulator

Let's start at Apple's iOS Dev Center—`https://developer.apple.com/devcenter/ios/index.action`. I'm going to assume that somebody intelligent enough would have purchased my book, and is fully capable of registering and creating an account. What next?

1. Download the latest SDK (4.3 at the time of writing) and the XCode and iOS package.

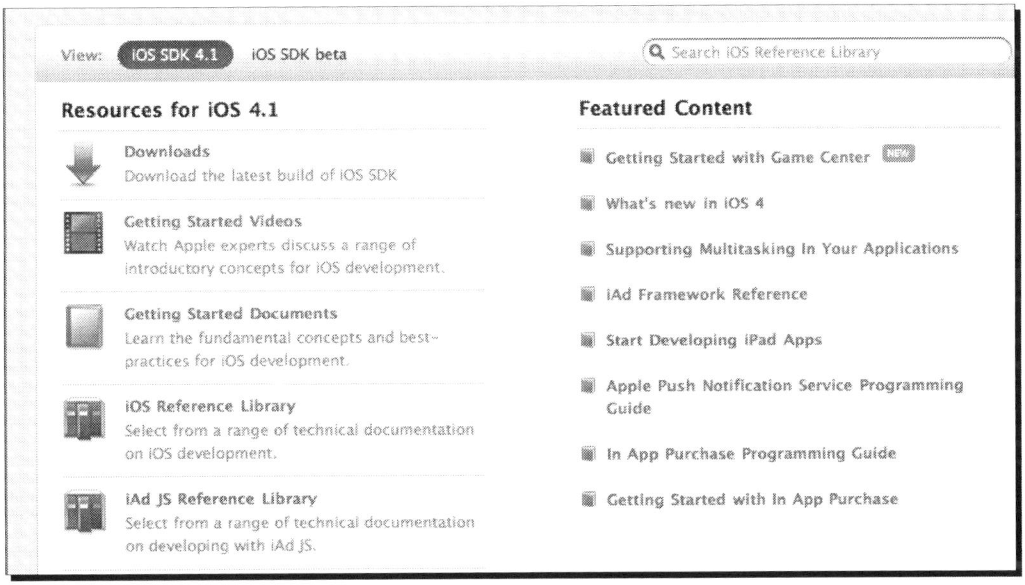

2. Wait for the three gigabyte download to finish. This may be a good time to get a cup of coffee.
3. Run the installer!

4. Launch Xcode, and start a **New Project**—select **iOS** and **View-based Application**. Name your application **FirstApp** when prompted.

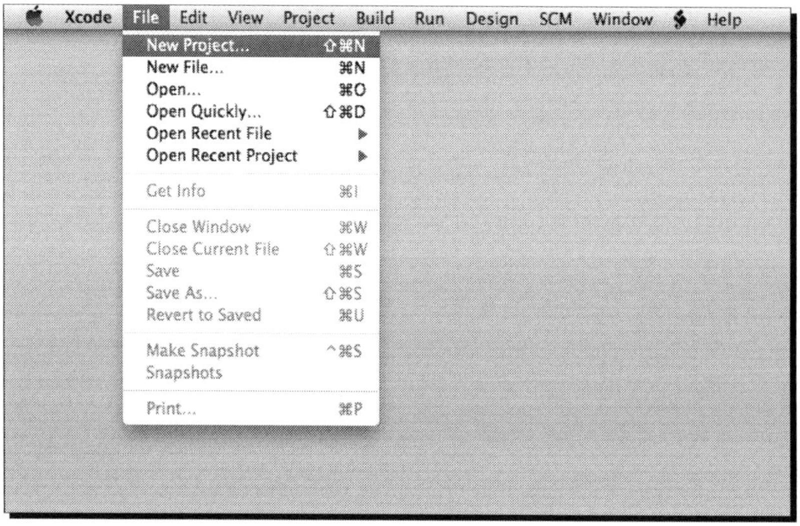

5. The following screenshot shows what the final application should look like in Xcode:

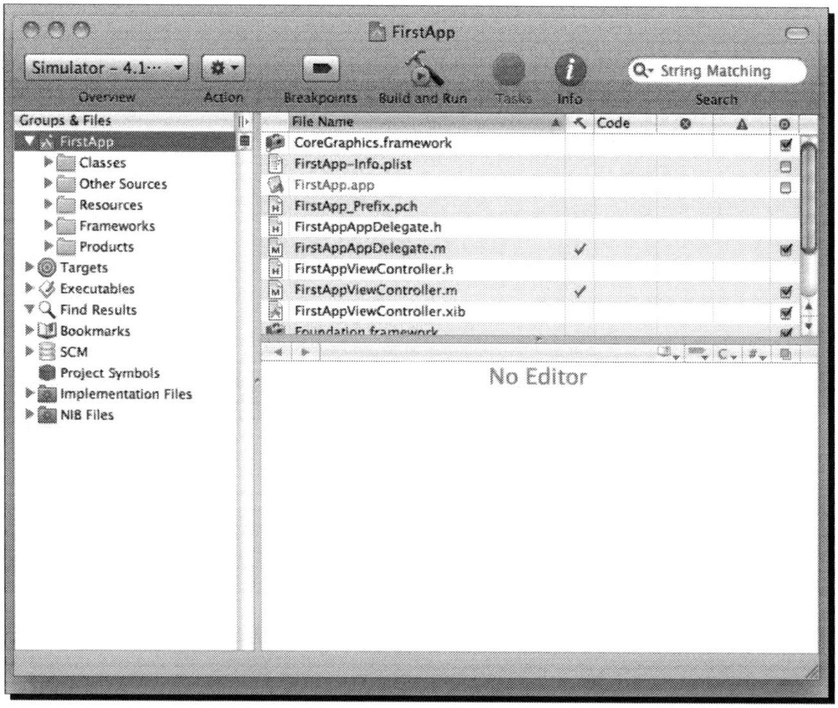

6. Hit **Build and Run** on the Menu Bar—you should see the iPhone simulator launch on your screen. If you go to the home screen on your simulator, you should see the icon for your application:

7. This verifies that your setup is ready to begin writing PhoneGap applications for iOS. Congratulations, you're an iOS developer!

What just happened?

We just wrote our first iOS application!

Okay, so we didn't actually write any code—perhaps the project title counts as code, but it's a bit of a stretch. But if you can get this far—installing Xcode and launching the iPhone simulator—then the rest of the setup should be fairly easy.

One part that can also be tricky is deploying your new application to a physical iOS device. Please check *Appendix, Deploying to an iOS Device*, for help with this; you can also consult Apple's documentation at `http://developer.apple.com`, for the most up to date details.

Installing PhoneGap-iPhone

First things first—yes, it should be called PhoneGap iOS. As the old programming saying goes, there are only two hard problems in Computer Science: cache invalidation, off by one error and naming things.

You'll notice that neither of those problems involved PhoneGap, which is a doddle by comparison. Let's get going.

Time for action – Hello World with PhoneGap-iPhone

1. Open your OS X **Terminal**, and navigate to a folder you don't mind writing to. Make sure you have the **Git** installed and available in your PATH.

2. Enter the following command:

```
$ git clone git://github.com/phonegap/phonegap-iphone.git
```

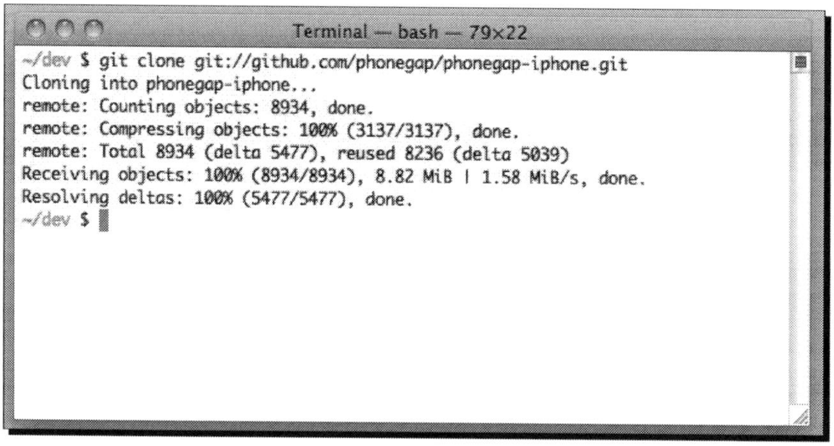

3. Now change into that directory, build the installer, and run it:

```
$ cd phonegap-iphone
$ make
$ open PhoneGapLibInstaller.pkg
```

4. You'll see the PhoneGap GUI installer in front of you. Follow the instructions onscreen—the installation process takes up less than a megabyte on disk and doesn't require administrative privileges, so you shouldn't encounter any errors.

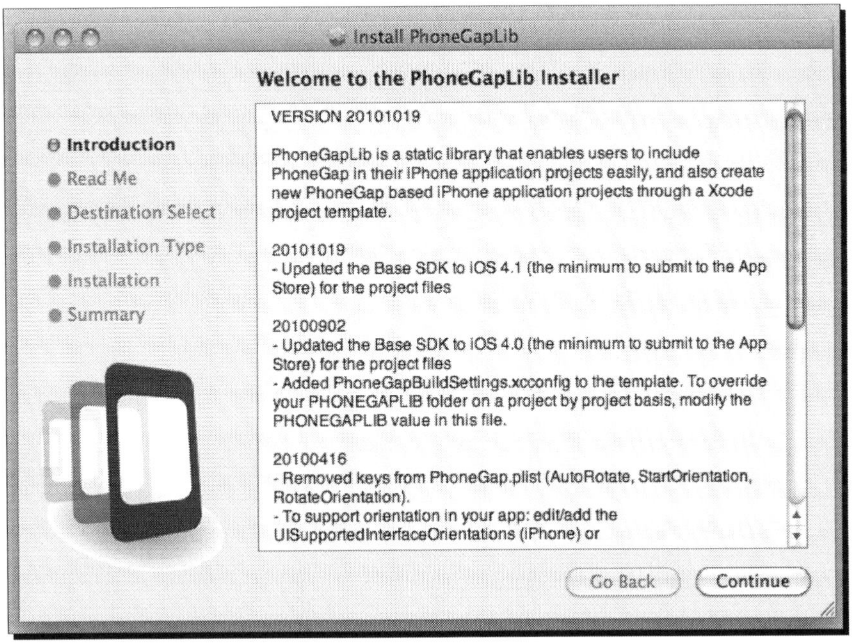

5. Quit and reopen Xcode if you still have it open—it will need to be restarted for the PhoneGap Project Template to be visible.

4. From the newly opened Xcode, select **New Project** again, and this time choose **PhoneGap** from the **User Templates** section, and **PhoneGap-based Application** from the main pane. Call your new project **FirstGapApp** (or, you know, something clever). You should see the familiar project view on Xcode, along with a `www` directory on the left-hand side.

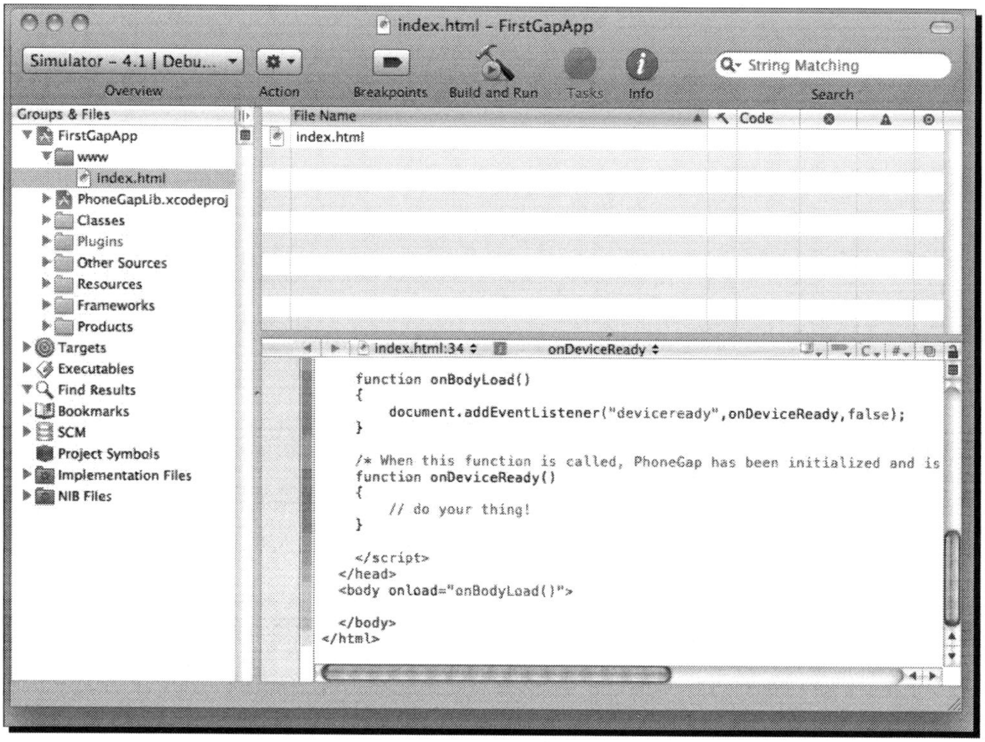

5. Open `www/index.html` from the left pane of your application window. Scroll down to the JavaScript function `onDeviceReady` and add the following line of code:

```
alert('Hello PhoneGap!');
```

6. Hit **Build and Run** to see the results.

7. Your JavaScript code has now executed, and your PhoneGap application is ready to go.

What just happened?

We got our first PhoneGap application up and running, that's what—an application running natively on a mobile platform that is wholly controlled through HTML, JavaScript, and CSS. We should give ourselves a pat on the back for that alone!

We can get an initial sense of how PhoneGap works if we look at the left-hand side of the Xcode window, and contrast it with what we saw on FirstApp. Here are the important things to note:

- There is a blue folder called `www`, next to a bunch of yellow folders similar to those in FirstApp. In Xcode's world, a blue folder is a directory on the file system that is bundled in with your project, whereas yellow folders are virtual directories containing project source code.

  It's important to be aware of this for one reason especially—as bundled files, PhoneGap source files are not automatically refreshed on each compile of your application. If you want to refresh your application in the simulator, or on your device, you will need to **Clean** it first.

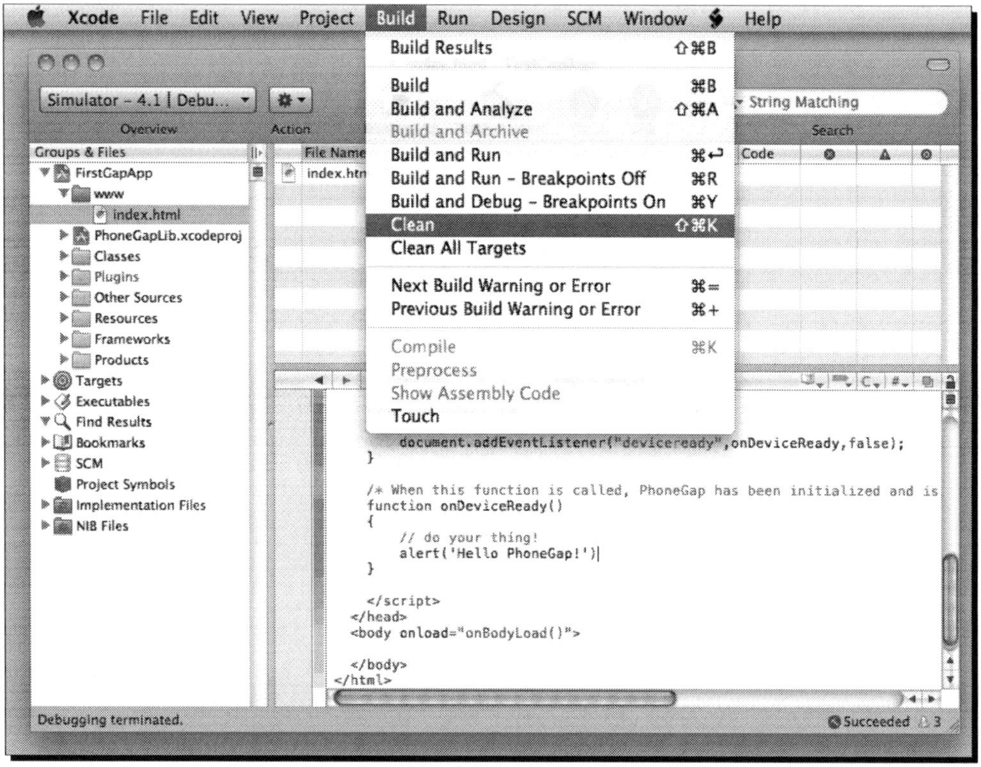

- There is a second blue Xcode project called **PhoneGapLib.xcodeproj**, below the main **FirstGapApp** Xcode project. This is the static PhoneGap library that was installed by the installer, and that GapFirstApp links to. If you wish, you can double-click on the project and edit away at the PhoneGap library—that's the beauty of open source software. But don't do that just yet.

- The more eagle eyed among you will have noticed that the www folder contains only `index.html`, but it requires a file called `phonegap.js` on line 15. This JavaScript file isn't strictly necessary for PhoneGap development, but it does give you access to all of the PhoneGap APIs. By default, it's autogenerated when you build your application.

There are a couple of other differences between a PhoneGap-based Xcode application and a regular view-based iOS application, but we'll come to those in due time. Let's play around with our iPhone application for a bit, and then move onto the next platform.

Pop quiz – PhoneGap iPhone basics

1. Where are your PhoneGap assets (HTML, JavaScript, and CSS) located in an Xcode project?
    a. In the project root
    b. In the `phonegap` folder
    c. In the www folder
2. How do you rename a PhoneGap iOS application?
    a. Change the `<title>` tag in `index.html`
    b. Rename `index.html` to `SomeThingNew.html`
    c. Edit the `Application-Info.plist` file
3. What function is called by the `alert('Hello PhoneGap!')` code?
    a. The standard `alert` function in the iOS WebView
    b. The `alert` function defined in `phonegap.js`
    c. The native Objective-C notification API that PhoneGap links to

Getting started with Android

Google's Android operating system is, in many ways, the antithesis of iOS: open instead of closed, and fragmented instead of integrated. This applies to the development environment as well—Android is a less bureaucratic environment than iOS, but has a few more rough edges along the way.

A note on development environments

It was the Roman playwright Terence, of the second century BC, who wrote *Homo sum, humani nihil a me alienum puto*; I am human, nothing human is alien to me. I feel likewise, except where the **Eclipse** IDE is concerned.

There are Eclipse plugin for both Android and BlackBerry development, which are certainly compatible with using PhoneGap on each platform. However, the major benefit of these plugins is their assistance with Java development, which is not the chief concern for developers using PhoneGap. Any text editor is sufficient for developing HTML, JavaScript, and CSS.

Rest assured, if Eclipse is your preferred environment, all of the content for these two platforms also applies in Eclipse.

Time for action – Getting the SDK running

Android isn't tied to a single IDE in the manner iOS is tied to Xcode, although you can use the **ADT** plugin for Eclipse for a somewhat similar experience. There is also an Android plugin for IntelliJ IDEA available. For PhoneGap's purposes, this is a boon: we can go straight to a PhoneGap application, once the basic SDK is set up. Let's do that now, and then get on to the good stuff:

1. Download the latest SDK package (r11 at the time of writing) from `http://developer.android.com/sdk/index.html`.

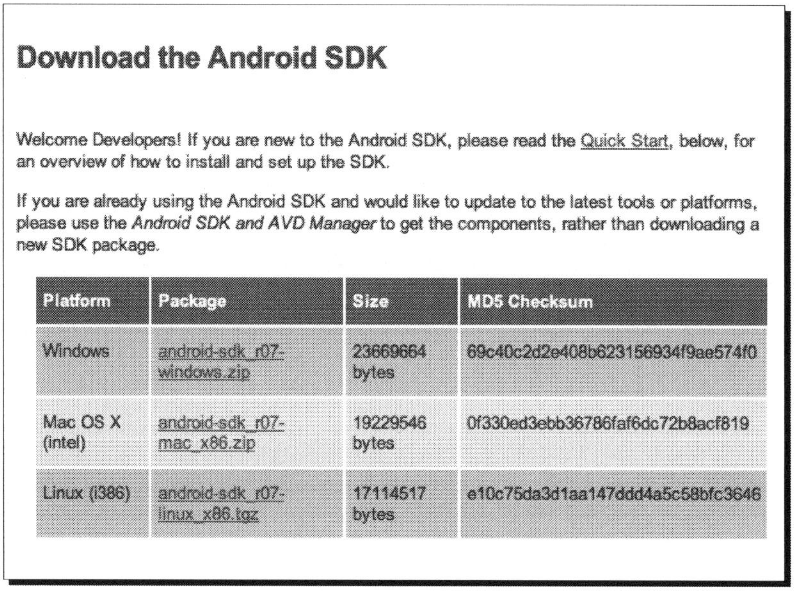

Download the Android SDK

Welcome Developers! If you are new to the Android SDK, please read the Quick Start, below, for an overview of how to install and set up the SDK.

If you are already using the Android SDK and would like to update to the latest tools or platforms, please use the *Android SDK and AVD Manager* to get the components, rather than downloading a new SDK package.

Platform	Package	Size	MD5 Checksum
Windows	android-sdk_r07-windows.zip	23669664 bytes	69c40c2d2e408b623156934f9ae574f0
Mac OS X (intel)	android-sdk_r07-mac_x86.zip	19229546 bytes	0f330ed3ebb36786faf6dc72b8acf819
Linux (i386)	android-sdk_r07-linux_x86.tgz	17114517 bytes	e10c75da3d1aa147ddd4a5c58bfc3646

2. Unpack the contents of the SDK to a safe location, and then add that location to your operating system's PATH environment variable.
3. Launch the **SDK and AVD manager**—on Windows, run the `SDK Setup.exe` program, otherwise enter `android` at your prompt.

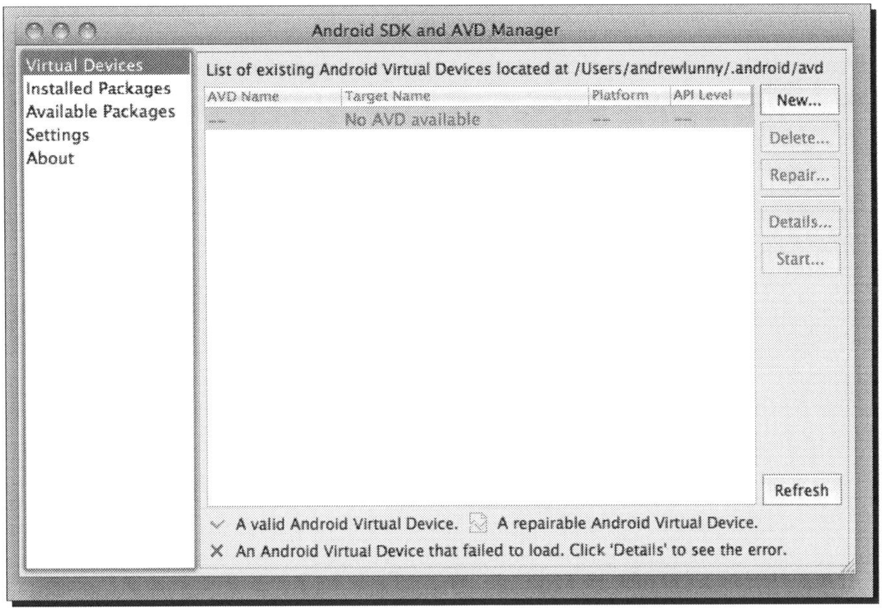

4. Install some packages! Select **Available Packages** on the left, and click everything that looks good. In particular, you'll want the **SDK Platform**, **Google APIs**, and **Documentation** for the version of the Android API you want to target and the latest **API tools**. I would recommend installing 2.2, with APIs 8: the emulator for 2.3 has a bug with PhoneGap, and later releases are targeting tablets, primarily. Be forewarned: this will take a while to download.

5. Now we need to create an **AVD**. Unlike iOS, there isn't a default simulator provided with the SDK, so we need to create one (there are lots of reasons for this, but no good ones). Select **Virtual Devices** on the left, then press **New...** on the right.

6. The settings are up to you, but a few recommendations:

   Go for the latest release of the Google APIs. Google APIs offer access to proprietary services offered by Google, in addition to the open source Android APIs. Most consumer devices will have these APIs available. The one major vendor that does not use Google's APIs is China Mobile—if you're developing for the Chinese market, just use the stock Android APIs. WVGA800 (480 pixels by 800) is a good resolution to go with.

7. The SD card should have at least a few hundred megabytes—this should be available to you on the menu shown below:

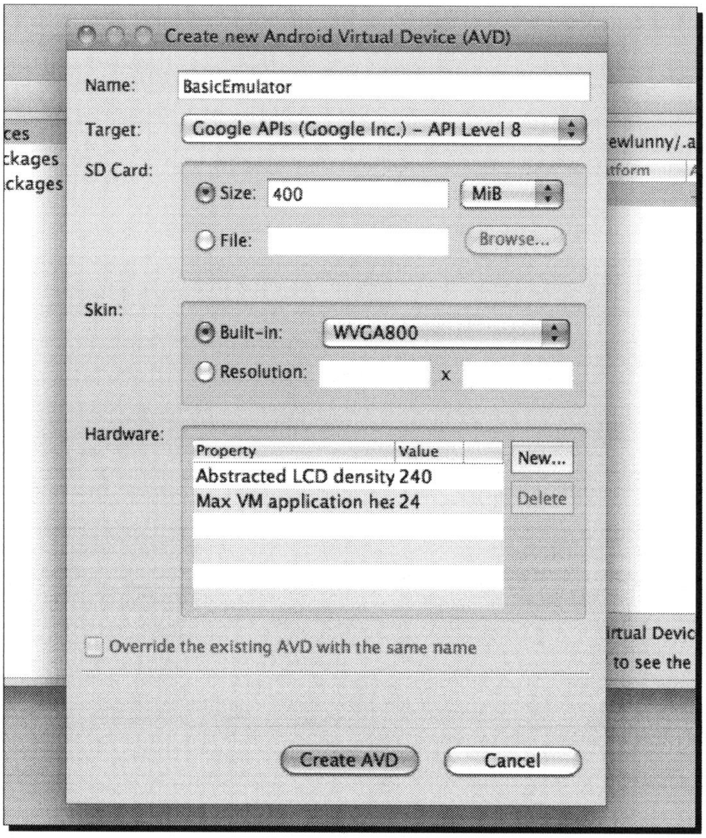

7. Select your new AVD from the list and hit **Start** (if the screen looks funny, adjust the screen size when prompted—I find around 8-10 inches gets a good size). It will take a while to boot up, during which time you can read the OS's source code, and bask in Google's openness/beneficence/omniscience.

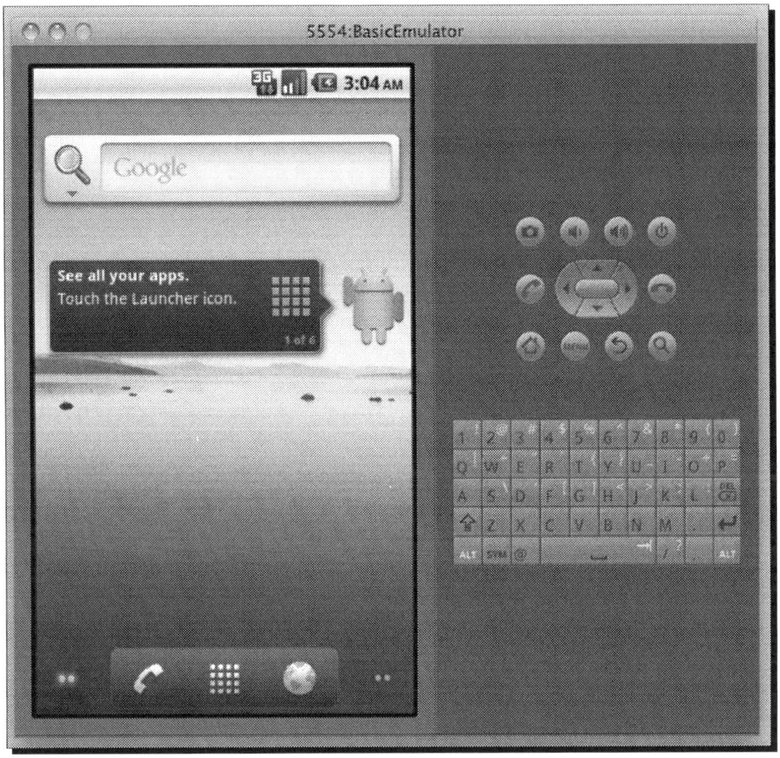

It may not be pretty, but you can read the source.

What just happened?

It's a bit more subtle than iOS, but we did more or less the same thing as the first tutorial of the chapter—got the SDK up and running, and launched an emulator.

One thing that's immediately apparent is that targeting the Android platform is not the same thing as targeting the iPhone device, or even the iPhone, iPod Touch, and iPad family of devices. There isn't a uniform screen resolution, operating system revision, or amount of storage space we can rely on, and the SDK tools don't allow us to easily build an emulator to represent, say, the HTC Desire phone. We need to use our own judgment to figure out the best emulator for our given application.

More than anything, we want to run applications directly on devices, to get the best sense of how they work. Luckily, this is easier on Android than on either of the other major platforms we will cover in depth. To allow app deployment via USB:

- Open the top-level **Settings** on your Android device
- Select **Applications**
- Check **Unknown sources**, to install non-Market applications
- Select **Development**
- Check **USB debugging**

If you're going to do serious Android development, we highly recommend getting a developer phone directly from Google—see `http://google.com/phone` for available devices. I recommend HTC's Nexus one, if it is available in your region. You can also purchase an Android Dev Phone from the Android Market—`http://market.android.com`—though you will need a developer account, which costs $25 US. The Android emulator experience is frustrating enough, and the deployment to phone process is smooth enough, to make this the outstanding approach.

Of course, we don't have any applications yet, but that should change shortly.

PhoneGap Android

Android is the second most mature PhoneGap platform, after PhoneGap-iPhone, and the development process is highly polished at this point. To avoid confusion, we should emphasize the difference between the two related, but distinct, parts of PhoneGap Android:

- **The PhoneGap Library**, or `phonegap.jar`: The Java library that links the PhoneGap APIs into the Android WebView, and initiates an app with that WebView
- **Droidgap**: A Ruby-based generator/utility for creating and deploying PhoneGap Android projects

Droidgap was created, at least partially, from frustration with the Android development process. It attempts to streamline and enhance a lot of the long-winded steps of Android development. However, it's a little brittle at the time of writing, so your patience is appreciated.

Time for action – Hello World on PhoneGap Android

1. Firstly, ensure that you have all the dependencies set up: **ant**, **git**, and **Ruby** should all be available on your path. Please see earlier in this chapter for help getting these set up.

2. Let's start like phonegap-iphone, with a git clone into somewhere sensible:

   ```
   $ git clone git://github.com/phonegap/phonegap-android.git
   ```

3. The next bit is a bit tricky, so bear with me. Ensure the Ruby executable `bin/droidgap` is somewhere you can access, either in your `PATH` or somewhere you can access directly (as in the screenshot below).

4. Switch to the `example` directory, run `droidgap create`, then switch to the `example_android` directory (again, I'm relying on the screenshot to make sense of this):

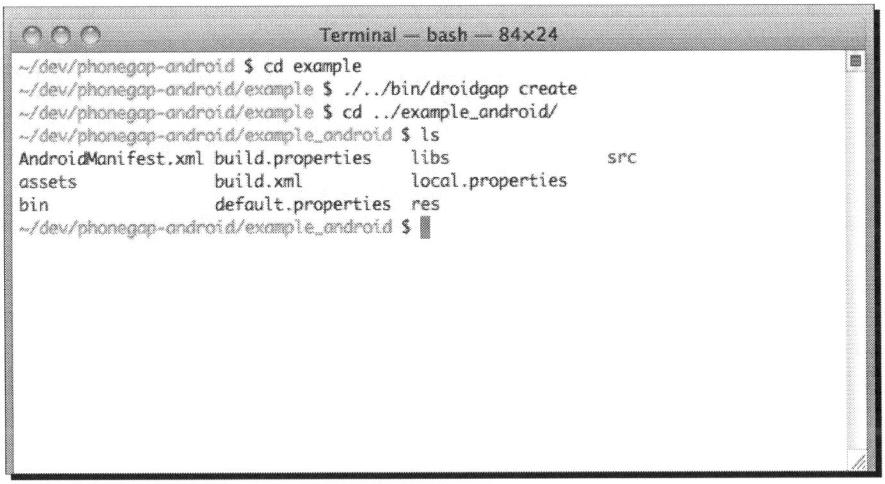

5. Build and install the application (this requires either a running simulator or an attached device):

   ```
   $ ant debug install
   ```

6. Pull up your AVD/non-virtual device, and check out your first PhoneGap Android project!

What just happened?

Well, the not-quite-as-elegant-as-we-had-hoped command line tooling for PhoneGap Android has generated a sample application that demos all of the PhoneGap functionality.

The `droidgap create` command is, currently, the smoothest way to create a sample PhoneGap Android application from a given set of client-side assets. We can then use the predefined ant tasks to build the application itself and install directly to our virtual or physical device.

The PhoneGap Android tools allow you, as a developer, to be further removed from the nitty gritty of the Java implementation and focus on the client-side technology that counts.

What's in a PhoneGap Android application, anyway?

One side effect of the droidgap app creation process is that it's possible to miss the structure of the application itself, which you can't really do with iOS development through Xcode. Here's what the project contents look like:

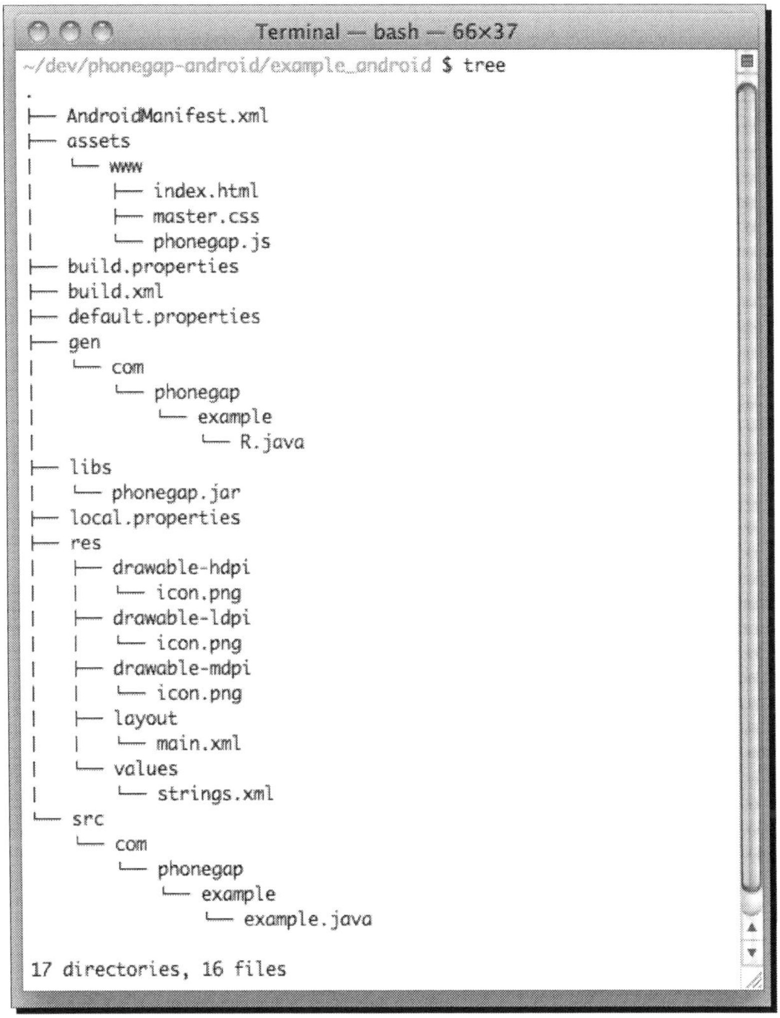

```
Terminal — bash — 66×37
~/dev/phonegap-android/example_android $ tree
.
├── AndroidManifest.xml
├── assets
│   └── www
│       ├── index.html
│       ├── master.css
│       └── phonegap.js
├── build.properties
├── build.xml
├── default.properties
├── gen
│   └── com
│       └── phonegap
│           └── example
│               └── R.java
├── libs
│   └── phonegap.jar
├── local.properties
├── res
│   ├── drawable-hdpi
│   │   └── icon.png
│   ├── drawable-ldpi
│   │   └── icon.png
│   ├── drawable-mdpi
│   │   └── icon.png
│   ├── layout
│   │   └── main.xml
│   └── values
│       └── strings.xml
└── src
    └── com
        └── phonegap
            └── example
                └── example.java

17 directories, 16 files
```

Some parts of note:

- `AndroidManifest.xml` is the equivalent of `Application-Info.plist` on PhoneGap iPhone—global settings, like the package name of your app, are put here, and in `res/values/strings.xml`.

- Your PhoneGap code is in `assets/www`, not just `www`.
- There are three copies of `icon.png` in the `res` folder, based on the DPI of the target screen (this is in common with iOS, just with a different directory structure). Droidgap just copies the same file in each location, but you can change this when getting ready to submit your application to the Android Market.
- Since it's Java at heart, the qualified name of your app (in this case, `com.phonegap.example`) has to be the name of your main Android activity. It falls under "tedium we try to avoid" but you should be aware that your AndroidManifest has to match your directory structure, or bad things start to happen.

Have a go hero – going further with Android

Check out the other Android command line tools—in particular, `adb logcat` and `adb shell`. Obviously Android has a lot more of the OS open than iOS, but what are some of the cool things you can do with that?

If you're familiar with the Ruby programming language, look into the source for `droidgap` (in the `lib` directory of the phonegap-android git repository). See if you can figure out what exactly it's doing, and how it works.

Getting started with BlackBerry web works

On to BlackBerry, the oldest of the three mobile platforms we're focusing on and the one with the most baggage.

In the annals of PhoneGap lore (two months ago), PhoneGap supported BlackBerry through the BlackBerry **JDE Component Pack**, on devices running RIM's BlackBerry OS 4.2 and above. While the code is still available, it has since been deprecated. It was, frankly, a nightmare, both in terms of PhoneGap's implementation and the browser that was ultimately exposed. You can look at some of the older tutorials on the PhoneGap wiki for evidence of these horrors.

As of BlackBerry OS 5.0, there is a better way—the **BlackBerry Web Works SDK**. Based on the W3C's web widget specification, which dovetails very nicely with PhoneGap, BlackBerry Web Works move PhoneGap BlackBerry from the slowest major PhoneGap platform to one of the very fastest.

As with every platform, and with web development in general, you still need to tread with caution regarding JavaScript and CSS support. In particular, if you do choose to work with devices running less than 5.0, the embedded browser is quite poor.

On top of that, the OS 6.0 devices (right now, just the BlackBerry Torch), have a credible, WebKit-based browser to hook into. Whatever next!

Time for action – Your first PhoneGap BlackBerry app

First things first: to run the BlackBerry simulator (as of the time of writing), you will need to be running on Windows—if you've been following from the start of the chapter, now is a good time to throw your Mac out the window. Or reboot into your other partition, whatever works.

1. Let's start by downloading the SDK: `http://na.blackberry.com/eng/developers/browserdev/widgetsdk.jsp`.
2. You're probably used to this bit by now, but here it goes: run the SDK installer. One thing to note is that you should install to `C:/`, rather than `C:/Program Files/`; this will avoid some issues with permission when running the ant scripts.
3. Clone the PhoneGap repo (in this case, we're using the `git bash` program to run git on Windows—other options are available):

```
$ git clone git://github.com/phonegap/phonegap-blackberry-
webworks.git
```

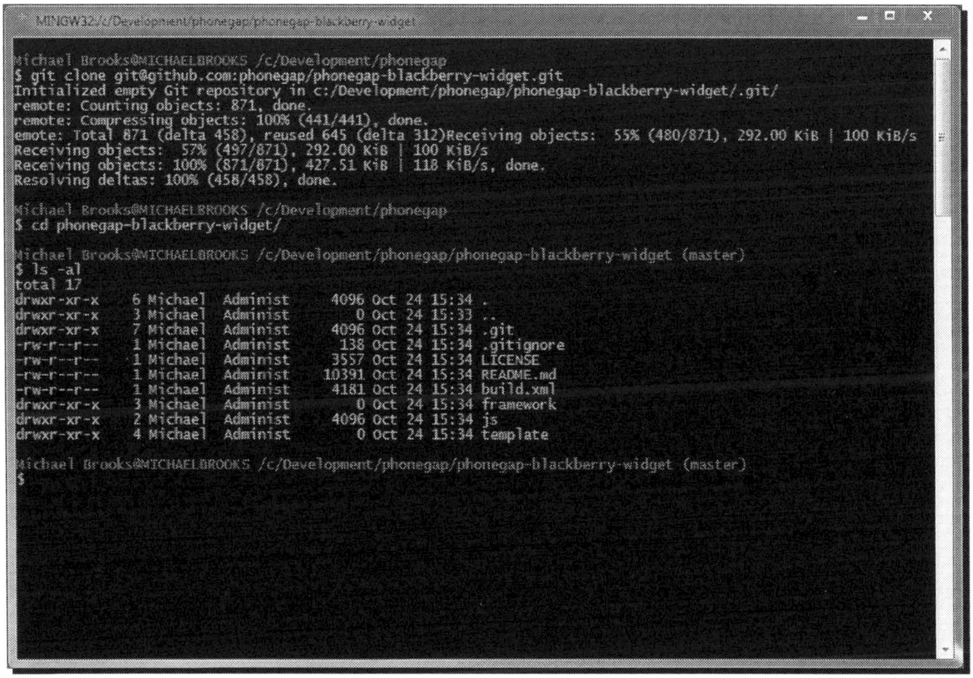

4. Like with PhoneGap Android, we'll use ant to generate our sample application:

```
$ ant create -Dproject.path="C:\Development\FirstApp"
```

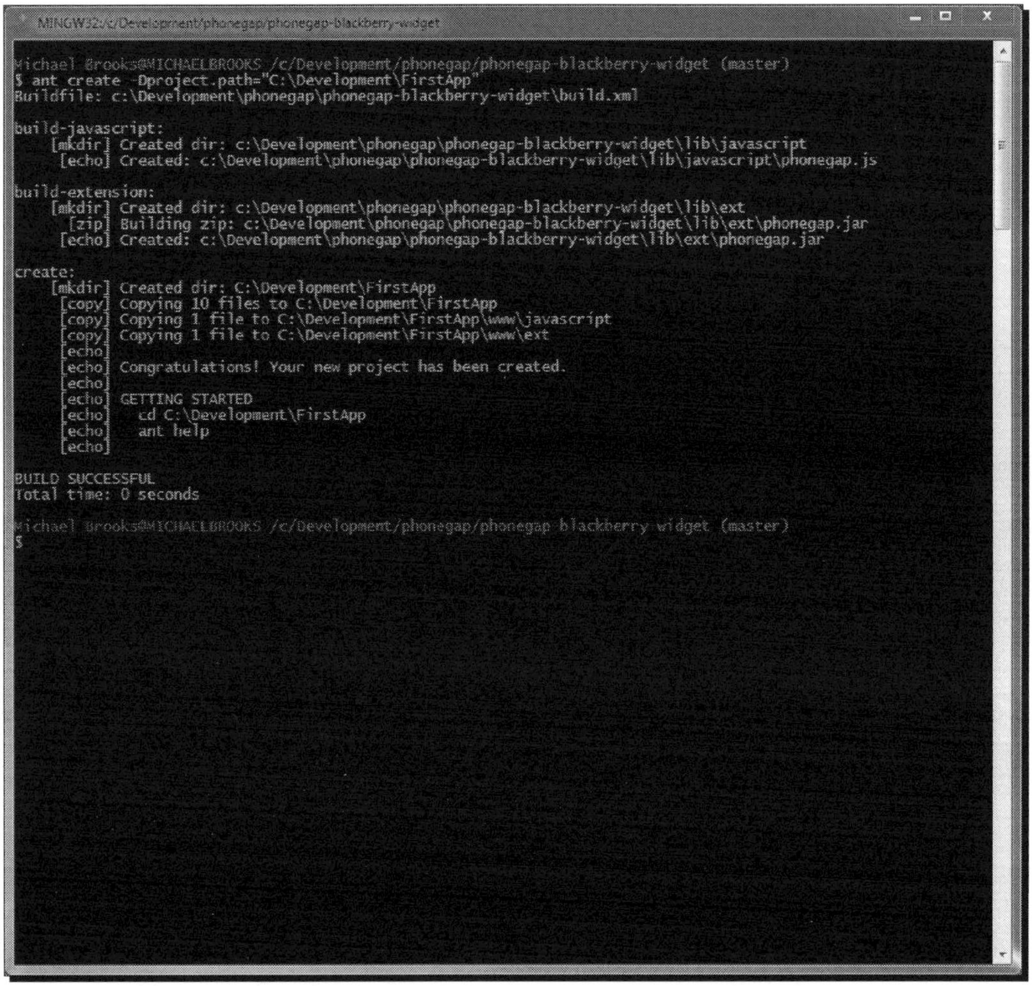

5. One quick gotcha is here: if you're running a 32-bit build of Windows, you will need to edit the `project.properties` file in your `FirstApp` directory, changing the `bbwp.dir` variable to point to where you have installed the SDK. The ant scripts can be a little bit brittle with older versions of Windows, so make sure that the WebWorks Packager is referenced correctly.

6. From your `FirstApp` directory, launch the simulator:

```
$ ant load-simulator
```

Or, if you're ambitious/determined enough to have your code signing set up already, launch your application on a device:

```
$ ant load-device
```

7. A couple of processes will be launched—the **BlackBerry simulator** and the **Mobile Data System Connection Server (MDSCS)**. The simulator is self explanatory, but the MDSCS is necessary for the simulator to server network requests.

I'm not sure exactly how much detail you'll want on this, but here goes: on a BlackBerry device, all network traffic is proxied through RIM's servers. On a BlackBerry simulator, everything is proxied through a local server. If you close that server, your application will silently fail, and an army of RIM ninjas will attack you while you sleep. Allegedly.

8. At any rate, the simulator is launched now. To find the PhoneGap Sample, hit the BlackBerry button (the one with seven circles resembling a **B**), then **Downloads**:

You should now see the icon for the sample application:

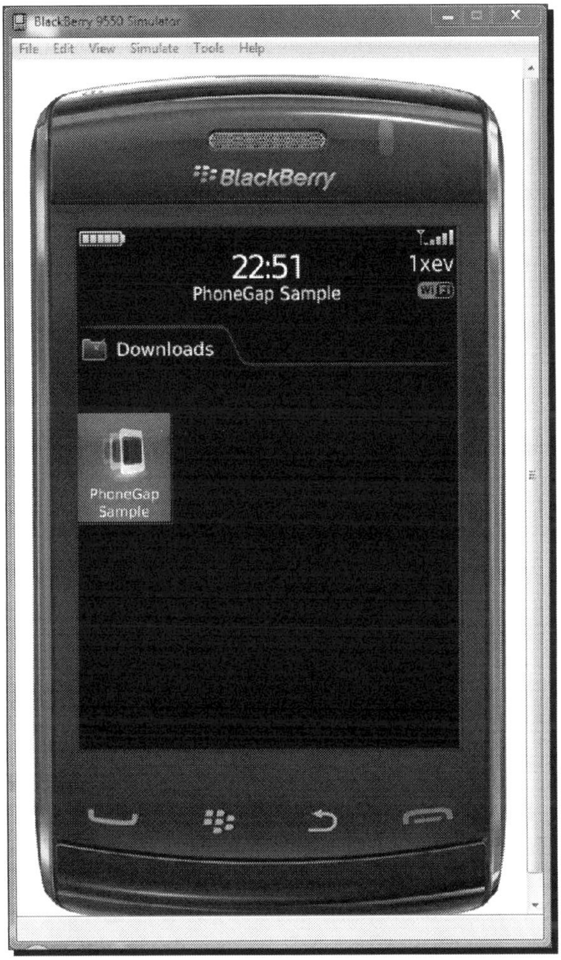

9. The next challenge is to launch the application. I'm sure you can figure this one out:

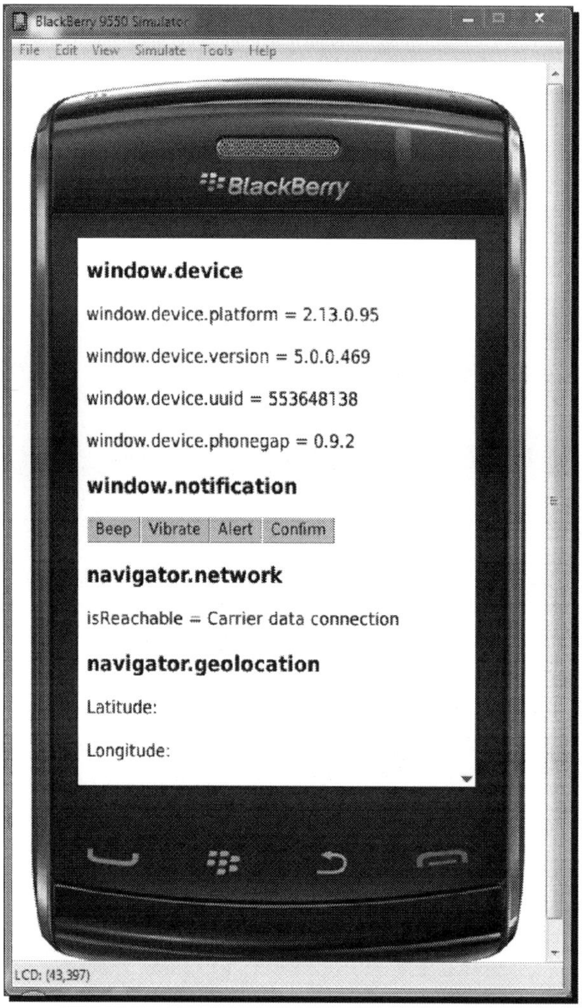

What just happened?

We got another application running on another mobile platform, using PhoneGap to fill in the gaps. Huzzah!

Things should be getting more familiar now—we use the PhoneGap tools to generate a sample project that contains a JavaScript reference to all of the PhoneGap APIs, and from then on we only need to worry about writing and rewriting the HTML, JavaScript, and CSS.

If we look at the structure of a BlackBerry Web Works PhoneGap application, things should look familiar:

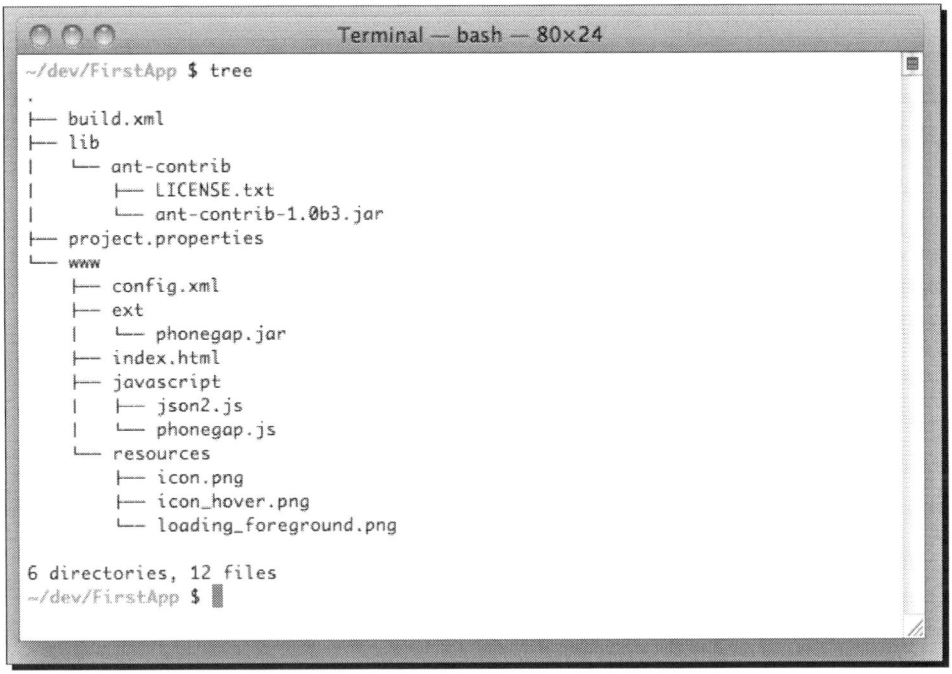

Once more, we have a `www` directory containing our `index.html` and `phonegap.js` files, a `build.xml` file (just like Android) for running ant tasks, an external library (`www/ext/phonegap.jar`) linking the native PhoneGap APIs, and an XML-based configuration file (`www/config.xml`) that defines global settings for our application.

Though there are major differences between the browser environments on each of these platforms, which we will cover in pain-staking detail over the rest of the book, a high-level look at PhoneGap on each of these platforms shows the commonalities.

Code signing for BlackBerry

Something I glossed over a bit earlier is Code Signing—like Apple with iOS, RIM requires you to sign your applications, with their approval, before you can install them on your (their) devices. Phew. It's an involved process, and we're all tired at this point, but here's an overview:

1. Go here: `https://www.blackberry.com/SignedKeys/`.
2. Fill out the form.
3. Allow RIM to relieve you of $20.
4. After a week or so, follow the instructions they email to you.

Say what you like about Apple, at least they're efficient at taking your money.

Have a go hero – Cross-platform fun

Already in this chapter, we've seen how PhoneGap gives similar structures to applications you can deploy to multiple mobile platforms. Try to take that a step further—can you link all three platforms to a single development directory? Do you need separate copies of `phonegap.js` for each platform?

Try to edit each platform's sample application. How do you remove, rebuild, and reinstall applications on each platform? Be sure to familiarize yourself with the documentation for each platform, in case any of these tasks become difficult—we'll try to take the SDKs for granted as we move on through the book.

Summary

Well that was a lot of fun (ahem). The mobile development landscape is strewn with clunky SDKs—if nothing else, the attrition that this chapter has covered should highlight the scope of the problem that PhoneGap aims to solve. From here on, things go a lot smoother.

Specifically, we covered:

- Getting an iOS PhoneGap application up and running
- Doing likewise for Android
- And once more for BlackBerry Webworks

We also discussed how you can begin to automate your workflow, using ant tasks where available. We're starting to look towards building cross-platform applications, rather than starting with a single platform and porting over—and now that we've installed all of these SDKs, we're going to spend as little time with them as possible.

Over the coming chapters, we will begin to add more and more functionality to our mobile applications, while building on the base we have established here. Our next step is to see how to structure mobile applications to get the best use out of PhoneGap.

2

Building and Debugging on Multiple Platforms

As I've stated a few times, the first chapter's work neatly covered the problem PhoneGap aims to solve: incompatible software and environments between different mobile platforms. From here on, things should be a lot more straightforward; we'll be developing with familiar, open web standards, and we'll start to actually write some applications.

In this chapter we will:

- Learn how to develop simple applications for mobile devices using a desktop browser
- Begin debugging JavaScript from within the browser
- Move our web browser-based project to mobile platforms
- Insert some native PhoneGap calls into our nascent application

So what application should we start with...

Designing with desktop browsers

So now we have an application concept—what's the next step?

Well we could break out PhoneGap and start native development straight away for a single platform—BlackBerry perhaps, since their users need the most encouragement. But this isn't a great approach—even using PhoneGap and simulators, as opposed to native code and physical devices, there's a large amount of overhead with native development. We need to build and compile our projects, install them on devices (virtual or physical), and then clean, uninstall, and rebuild whenever we make a small change. It's a good way to kill the early momentum of a project.

Because PhoneGap leverages open web standards (HTML, CSS, JavaScript), we don't need to go straight to the native SDKs—we can start work in a desktop browser, and then move on to a native project once the functionality is fleshed out. This way, we speed up our development cycles and spend more time implementing core functionality.

WebKit

Strictly speaking, you can use any major desktop browser to get started on your PhoneGap app—**Internet Explorer (IE)**, **FireFox**, **Safari**, **Google Chrome**, or **Opera**. All of these browsers (at least, the latest revisions) have adequate, at least, developer tools for logging and debugging our code.

For our purposes, however, we strongly advocate using a **WebKit** based browser—of those listed above, either Safari or Chrome. With the exception of Microsoft, all of the major mobile OS vendors now use a WebKit-based browser. This doesn't mean that a web application in Safari on OS X or Windows will behave just as it does on a Nokia N8 WebKit browser, for instance—but it will be a lot closer than running that same application in Opera.

If you're using PhoneGap just to develop the Windows Mobile and Windows Phone platforms, you would want to use IE in this part of the process.

Developing our first application: You Are The Best

We want to start on a positive note, so the project we'll develop in this chapter will be called **You Are The Best**, which will notify the user that they are the best. As people like to feel significant, our application should have a target market of six billion or so. As a developer, you can execute this application flawlessly, being, yourself, the best (see what I did there?).

That's our simple specification—produce a notification to the user on demand.

Time for action – Initial design and functionality

We remember from *Chapter 1, Installing PhoneGap* that PhoneGap apps are organized inside a www directory, with an `index.html` file as the starting point, so:

1. Create a directory for our project—`you_are_the_best`—and inside there, create a www directory with an empty file called `index.html`.
2. Fill out `index.html` with some HTML boilerplate and a positive initial message:

```
<html>
  <head>
    <title>You Are The Best</title>
   </head>
  <body>
    You're really great!
  </body>
</html>
```

3. Let's open that in Safari and see how it looks:

3. You'll notice I've resized Safari so it's fairly close to a smartphone sized-screen—it's an obvious approach, but very useful.
4. We have something, but not much. Let's put a few styles in there so it looks a bit more colorful—insert this `<style>` element into the `<head>` of your document:

```
<style>
  body {
    background: #ff0;
  }
```

```
    div#main {
      background: #ccf;
      height: 460px;
      width: 320px;
      padding: 10px;
      -webkit-border-radius: 8px;
  }
  </style>
```

5. That last property may well look a little strange—don't worry, we'll cover it later in this chapter.

6. Now modify the `<body>` element so it looks like this:

```
<body>
  <div id="main">
    You're really great!
  </div>
</body>
```

7. Save the file and reload Safari to see what we have now (please note the colors will appear differently on your full color screen than on this monochrome page):

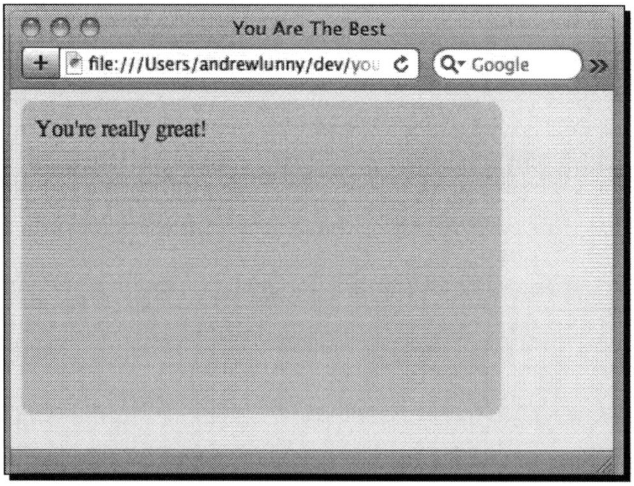

8. Well that certainly looks more colorful, but a bit of interaction would be nice—we want to tell the user directly that they are the best. Luckily, JavaScript has a built-in `alert()` function that does just this, and is never annoying to the user. Let's add a `script` element to our `head` to call `alert()`:

```
<script>
  document.addEventListener("DOMContentLoaded", function () {
```

```
    alert("We really do think that you are the best.")
  }, false);
</script>
```

Don't worry if that doesn't all make sense right now—we'll cover it later in detail.

9. Once more, save and reload Safari—you'll notice the alert pop up loud and clear:

10. This is great! But we would like to be able to see this message on demand, whenever we want. Let's write a JavaScript function to display the alerted message (this goes in the same `script` element):

```
function alertCompliment() {
  alert("We really do think that you are the best");
}
```

11. We'll add a button to the `div#main` tag to show the alert:

```
<div id="main">
  You're really great!<br>
  <button id="likeMe">Tell me I'm great</button>
</div>
```

12. Then attach the `alertCompliment()` function to that button (in the `script` tag, in the `DOMContentLoaded` listener):

```
document.addEventListener("DOMContentLoaded", function () {
  alert("We really do think that you are the best.");
        document.getElementById("likeMe").
addEventListener("click", alertCompliment, false);
}, false);
```

Again, if you're not sure why that code went in that place, don't worry. We'll get to that.

13. Save and reload. Now try clicking the button.

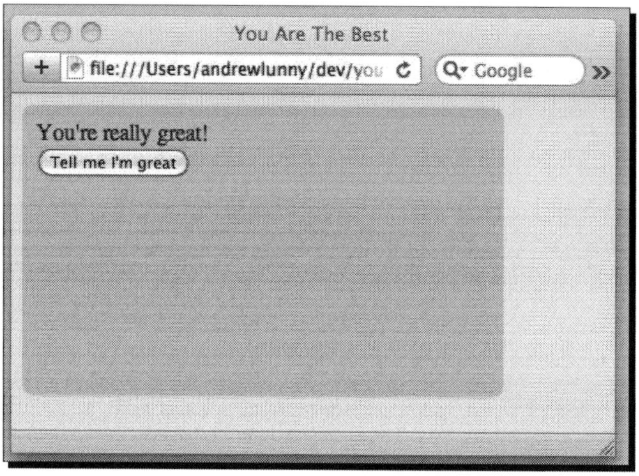

14. Voila! Now, to the App Store!

What just happened?

Alright, **You Are The Best** is no **Plants versus Zombies**—it's barely an **iFart**—but we've managed to get some basic functionality and styling into an HTML-based application with very little code and very little effort. These are the fundamentals of development with PhoneGap—rapid development in a stable desktop environment, quick iterations, and unobtrusive CSS, and JavaScript.

Our workflow

After the fairly exhausting bureaucracy of the first chapter, it will have come as some relief to avoid generating projects, building projects, launching emulators, signing binaries, cleaning builds and so forth, ad nauseum. Of course, as your application takes on extra complexity, more involved logic will require spending more time on application behavior with specific devices; however, your initial goal should be to spend most of your time in a quick edit-save-reload loop, switching between your text editor and your web browser.

You will also notice that we are working locally—all of the Safari screenshots above point to a `file://` URL. This is the same protocol that all PhoneGap apps run from on the supported platforms—the web assets are served directly from the file system, and, in most cases, a remote server is only used to serve dynamic data that cannot be bundled at compile time.

There are lots of benefits to this approach—your application functions offline, there is lower latency for retrieving assets, and available bandwidth is less of a concern—but, at this stage, the main benefit is that you won't need to upload your files anywhere to get the full testing experience. Just open `index.html` directly from your file system to get the optimal PhoneGap development workflow.

Our styles

Cascading Style Sheets (**CSS**) are the standard approach to styling web pages and applications, and PhoneGap doesn't modify CSS rendering in any way. However, there are a few features of the CSS we wrote that may be unfamiliar:

Unobtrusiveness

All of our CSS rules for **You Are The Best** are kept in a single `<style>` element in the head of the document. If you've used CSS before, you may have applied rules directly onto an element, like so:

```
<p style="color:red">This is some red text</p>
```

That's something we want to avoid. There's a philosophical reason to avoid setting styles through the style attribute: we should endeavor to separate markup and presentation code, so that CSS rules are only in CSS files or style elements. For simple, small apps, such as the one we've just developed, style elements are sufficient; once we have more complexity, we will definitely keep our styles in external CSS files. Then the body of the HTML document just contains document markup.

More convincing than the philosophical argument is the pragmatic one: changing styles is more work if elements are styled individually. With a small document, it doesn't make much difference, but if we have 500 paragraph tags with a style attribute coloring them red, that's 500 changes we have to make. If each tag has a class attribute set, and the class sets the color to red, we only need to change the code in one place to change all of the paragraph colors.

Width and height

A confession—you don't want to set the width and height of your only top-level element, as we have done. Even if you just target a single platform, you can't be sure of every resolutions and DPI for every device on that platform, and setting an absolute size set will lead to problems down the road.

A more robust strategy is to use a combination of flexible widths and heights, CSS3 media queries, and the special `<meta name="viewport">` element. You'll see these features when we port our application to a PhoneGap project.

So why set width and height? Mainly, so we have a rough idea of what the screen size will actually be on our device—it's a useful way to help get a sense of proportions when developing on a desktop browser. Usually, this does not affect the appearance on a mobile device, but it will be worth keeping note of when deploying to our handsets.

-webkit-border-radius

Now here's something that's actually new! `-webkit-border-radius` is the first CSS3 property we'll encounter. It rounds the corners of a (rectangular) block element, as demonstrated in the following screenshot:

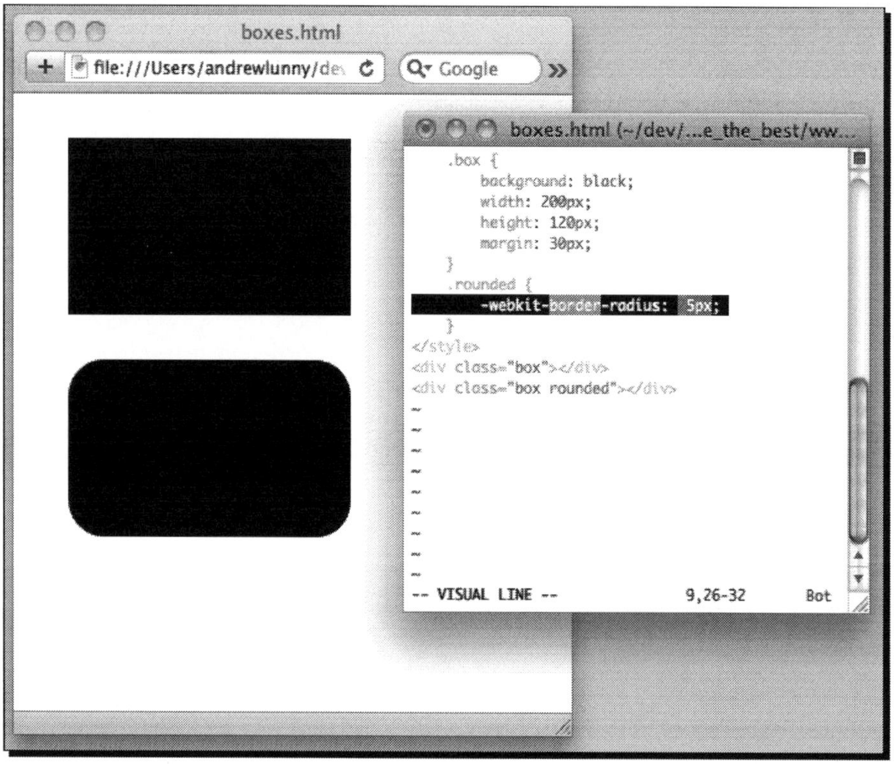

It's pretty neat, especially if you're familiar with the contortions web developers previously went through to achieve flexible rounded corners.

However, calling `-webkit-border-radius` a CSS3 property is a little inaccurate. The CSS3 property is `border-radius`, and `-webkit-border-radius` is a vendor-prefixed prerelease implementation of the border-radius spec. There's a similar `-moz-border-radius` property you would use when targeting FireFox.

`Border-radius` is actually one of the better implemented CSS3 properties, and we could have used the non-prefixed property for this example, but you should be familiar with these prefixes and what they mean:

- A `-webkit` prefixed CSS property will only work in some WebKit-based browsers—not older WebKit browsers, and not non-WebKit browsers
- Vendor prefixed CSS properties are not guaranteed to be spec-compliant
- New properties are initially only available in prefixed form

If you're writing a website for a general audience, you'll want to specify the property with every vendor's prefix, along with the non-prefixed CSS3 standard property, which is annoying but future-proof. For PhoneGap applications, you can be safe most of the time with using prefixed properties, but if the non-prefixed version is supported, that's the best bet to ensure your code is portable and plays nice with the specs.

Our scripts

The other notable (well, somewhat notable) part of **You Are The Best** was the JavaScript code. PhoneGap is at heart about writing mobile applications with JavaScript, and it pays to have a good understanding of the language. If you're looking for a reference guide to JavaScript in book form, check out David Flanagan's *JavaScript: The Definitive Guide*; if you prefer an online reference, your best bet is the Mozilla Developer Network at `http://developer.mozilla.org`, or Apple's Safari Developer Library, at `http://developer.apple.com/library/safari`.

As PhoneGap has a full JavaScript runtime on every platform, you are free to use a full-stack framework, such as Sencha Touch or jQuery Mobile. I would recommend using the standard JavaScript APIs to keep file size down, and to increase the clarity of your code. However, the benefits of these frameworks are well known. They're certainly popular options when combined with PhoneGap.

JavaScript is a beautiful, albeit often misunderstood language, and it's really worth your time to understand it's intricacies fully. Hopefully, once you've written a few PhoneGap applications, you'll get a good sense of the language's joys.

Unobtrusiveness

Much like our CSS, the JavaScript in **You Are The Best** is isolated to a single element in the document, rather than intruding on all of the **Document Object Model (DOM)**. In particular, our event handlers (the initial alert message once the DOM is ready and the subsequent alerts when our button is clicked) are not defined as attributes of the elements they are attached to. For many people, their first experience with JavaScript is writing code inline on the DOM, like this:

```
<button id="likeMe" onclick="alertCompliment(); return false">Tell me
I'm great</button>
```

Or even:

```
<a href="javascript:alertCompliment(); return false">Tell me
something, please!</a>
```

As with obtrusive CSS, this kind of code quickly becomes a maintenance nightmare—and for users with JavaScript disabled, a link like our second example will give a 404 HTTP error.

Since a mature PhoneGap application will involve hundreds of lines of JavaScript, it's important to keep our code well organized and robust from the start—and the best approach is to ensure JS stays only with JS. As with CSS, this may involve using an external file, or separate `script` tags: either way is preferable to using JavaScript inline in HTML attributes.

addEventListener

One of the means we have to ensure unobtrusive JavaScript is the `addEventListener` function, the DOM level 2 mechanism to register event listeners. The mature PhoneGap platforms we will be focusing on (iOS, Android, and the BlackBerry widgets runtime) all support `addEventListener`.

The ancestors of `addEventListener`, were, for DOM level 1, binding JavaScript to the `oneventname` property of the JavaScript object at hand:

```
window.onload = function () { alert("hello!") }
```

Or, for DOM level 0, binding through an attribute on the related HTML tag:

```
<body onload='alert("hello!")'> … </body>
```

`addEventListener` is the preferred mechanism for a couple of reasons:

- **Separation of Concerns**: With `addEventListener`, it's trivial to keep your event-handling JavaScript code separate from the code that renders your DOM.
- **Multiple Listeners**: With both the level 0 and 1 mechanisms, it's tricky to bind multiple handlers to a single event; with level 2, you simply call `addEventListener` multiple times.

- **Better Support for Custom Events**: For example, you can use `addEventListener` to listen for PhoneGap's custom `deviceready` event to fire—since the JavaScript `document` object doesn't have an `ondeviceready` property, older event binding methods cannot attach to this event.
- **Easy Unregistering**: You can use the corresponding `removeEventListener` to detach a listener from a specific event.

You should get used to `addEventListener`—it will treat you well.

DOMContentLoaded

You'll notice in the above examples the discouraged event examples both bound to the `onload` global event, on the `window` object or the `<body>` element. The `load` event does have its uses for code that should execute as soon as possible—in practical terms, anything that doesn't rely on the DOM being rendered. Our `alert` call in the sample code above could have been bound to the `load` event without any change to the user experience.

However, attaching to the `DOMContentLoaded` event is another example of a best practice that we should get the hang of early. This event, which is also supported in all OS the platforms we're interested in right now, fires after the page has fully rendered and the DOM is in a state with which we can confidently interact.

Pop quiz – Initial design

1. Let's say we wanted to add a second page to our application, with the same CSS and JavaScript code. What would be the best approach?
    a. Copy `index.html` to a new file, and then edit that new file
    b. Extract our CSS into an `app.css` file, and our JavaScript to an `app.js` file, and include those on the new page
    c. Create an empty `new_file.html` that inherits from `index.html`
2. **You Are The Best** has proved so popular that the users are demanding a tablet application release to go with their smartphone application. How should we edit the CSS?
    a. Make the absolute measurements (height and width) relative, so that they scale up to the size of larger devices
    b. Don't change anything—let the browser resize things natively
    c. Copy our existing CSS code to a `best.phone.css` file and create a new `best.tablet.css` file; conditionally load the correct one on out device

3. The other demand we've received from users is for a second button, that says **Tell my cat she's great too**. This button should present a different message to the user. How should we best implement this code?

    a. Refactor our `alertCompliment` function to base its message on the object that calls the function, and attach that function to `button#likeCat`

    b. Write a new function `alertCatCompliment` that is bound to `button#likeCat`

    c. Copy and paste the existing `addEventListener` call, replacing the message contents and the ID of the button appropriately

Using the web inspector

So that's all well and good for a moronically simple application, but what about when we have more complex logic? We will need some kind of debugging tools. Luckily, Safari now has an exemplary set of JavaScript tools, centered around the Web Inspector, a useful hub for CSS, JavaScript, and DOM manipulation.

There are also debugging tools for other browsers: Google Chrome has the same Web Inspector, since it is a part of the WebKit codebase; Firefox has a very popular plugin called Firebug, which is the precursor of most other web debugging tools; Opera has a similar tool called Dragonfly; and IE has equivalent developer tools also. I'm focusing on Web Inspector through Safari since, as mentioned, it's the closest desktop environment to the mobile-WebKit devices we'll be deploying to.

Accessing web inspector

It's not on by default, but it is dead simple to get access to the Developer Tools in Safari. To do so, we just need to go into Safari's **Preferences...** panel and enable **Show Develop menu in menu bar**:

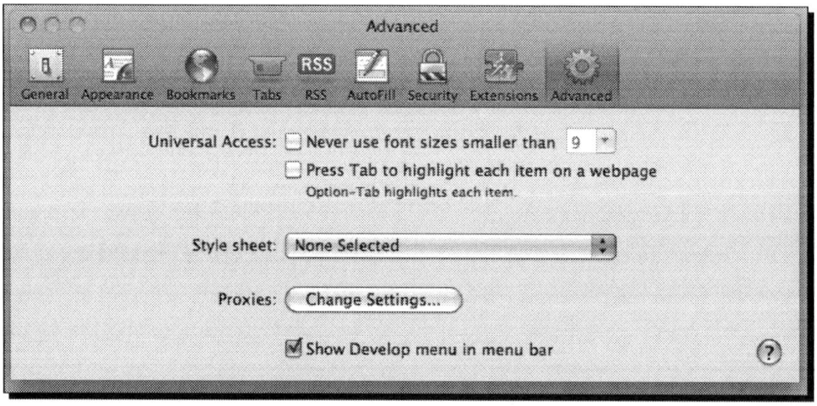

We can then display the inspector by choosing **Show Web Inspector** from the application's **Develop** menu:

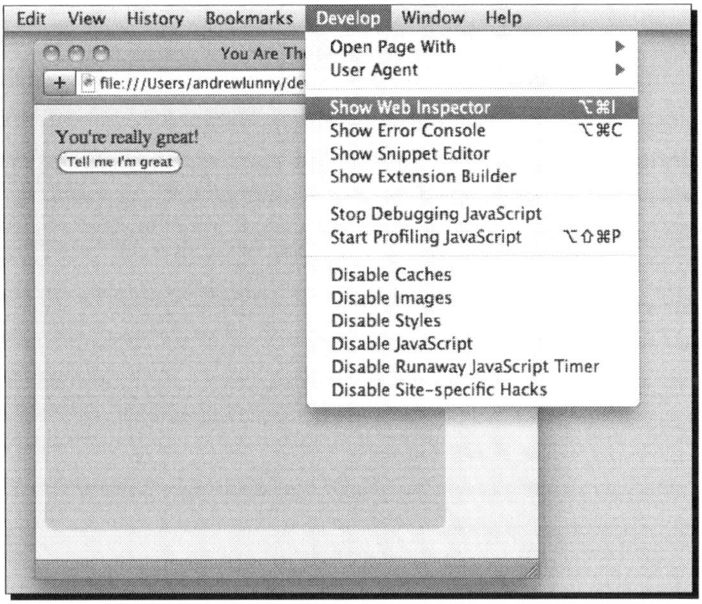

Time for action – Simple logging and error checking

Let's add a second button to **You Are The Best** with some new social behavior, and see what to do when things go wrong:

1. Open `index.html` in your text editor of choice, and add our second button immediately below the first:

```
<button id="likeMe">Tell me I'm great</button><br>
<button id="likeHer">Tell her I'm great</button>
```

2. Let's open the page in Safari to verify that it renders as expected:

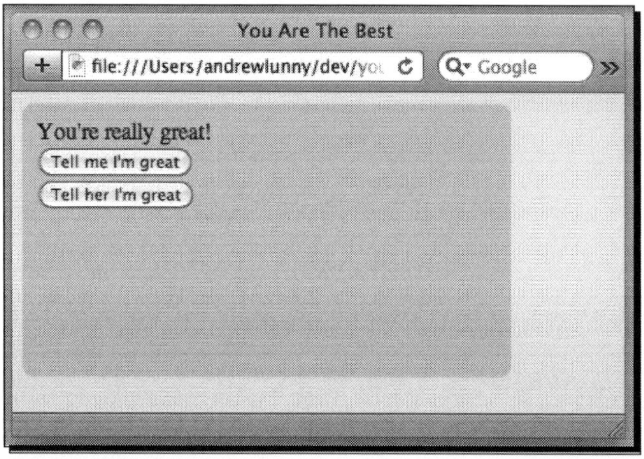

3. Now let's add some behavior for our new button. We'll rewrite the JavaScript as follows (changes are highlighted):

```
function alertCompliment() {
  var elementId = this.id,
  var greeting;

  if (elementId) {
    greeting = "you are";
  } else if (elementId == "likeHer") {
    greeting = "she is";
  }

  alert("We really do think that " + greeting + " the best.")
}

document.addEventListener("DOMContentLoaded", function () {
  alert("We really do think that you are the best.");

  document.getElementById("likeMe").addEventListener("click",
alertCompliment, false);
  document.getElementById("likeHer").addEventListener("click",
alertCompliment, false);
}, false);
```

4. Again, we'll refresh Safari, and this time click on our second button:

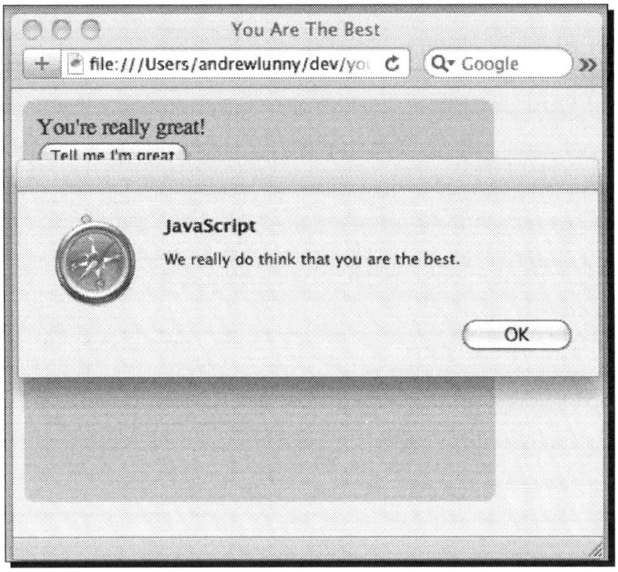

5. It is not what we expected. We see the first alert message no matter which button is pressed. Luckily, we can begin to use Web Inspector to see what's going on. We're going to add a few log calls to our `alertCompliment` function and see what's going on. We only have two variables, so let's verify the value of those during the function's execution:

```
function alertCompliment() {
  console.log("alertCompliment() called")
  var elementId = this.id,
  var greeting;

  console.log("elementId is", elementId)

  if (elementId) {
    greeting = "you are";
  } else if (elementId == "likeHer") {
    greeting = "she is";
  }

  console.log("greeting is", greeting)

  alert("We really do think that " + greeting + " the best.")
}
```

6. Now let's open the page in Safari again, but this time open Web Inspector also, press the first button once and dismiss the alert, and then press the second button and dismiss the alert. Now look at the output in Web Inspector's console:

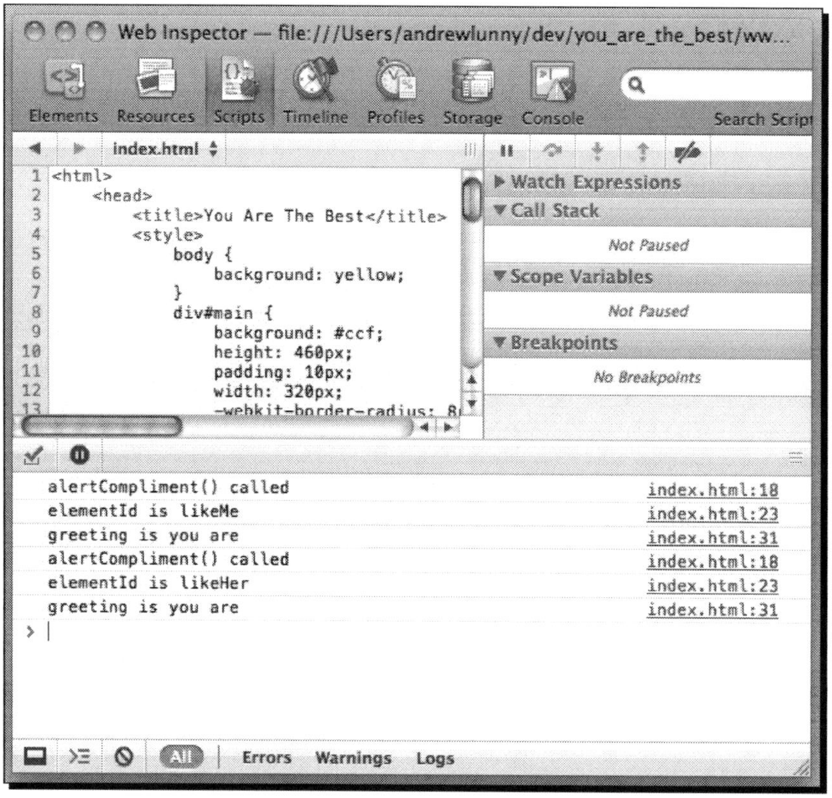

7. `elementId` is what we want, but `greeting` is not set correctly—the error is somewhere between our first and second `console.log` calls. Let's change the first `if` statement we have to be more careful:

```
if (elementId == "likeMe") {
  greeting = "you are";
} else if (elementId == "likeHer") {
  greeting = "she is";
}
```

8. Let's try the second button in the browser, aga... oh wow, it works!

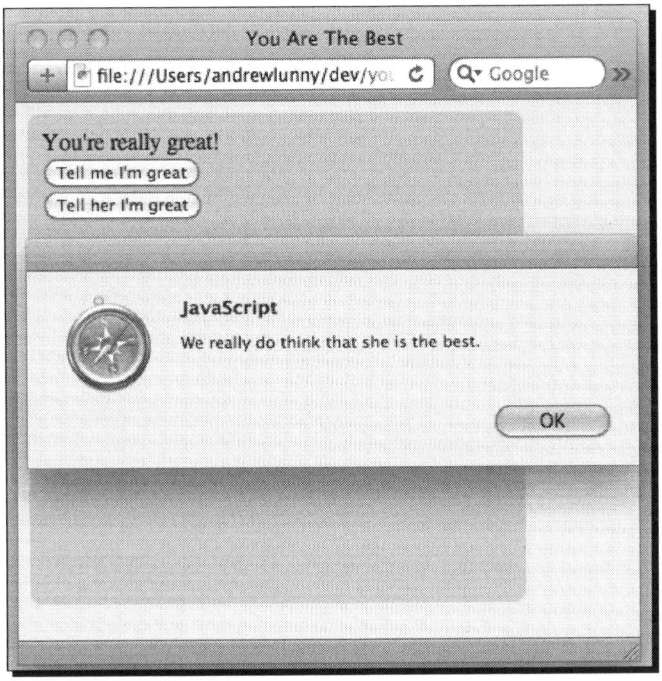

What just happened?

We got our first taste of Web Inspector, and the `console.log` function in particular. `console.log` and its aliased PhoneGap platform calls, is your run of the mill logger, but because JavaScript (in the browser) doesn't have a robust IO setup, as other languages do, the logger is always sent to a pre-determined view, usually determined by the browser. There's an easy way, for example, to redirect the output of `console.log` to a file, or to a separate logging service on your network.

Debugging is not a primary focus of the book, but if any of the code samples don't work for you, or work but don't work very well, it's a fine tool to have at hand. Your next step is to use `console.log` and the Web Inspector console to isolate what the problem is; once that's isolated, finding a solution should be trivial.

Have a go hero – Playing with Web Inspector and JavaScript

One of the neatest features of JavaScript is dynamic scope reassignment—that is, you can call a function in the context of any other function (and change what the `this` variable points to). Every function that is called has a magic variable, `this` that points to the outer scope of the function: either an object that it was called on, or the global scope. That's why our code works—the element `alertCompliment` was called on becomes `this`, which we can query for an `id`.

What exactly is going here? Look up `Function.prototype.call`, `Function.prototype.apply`, and `Function.prototype.bind`, and see what can be done with those tools.

We have also barely touched on the surface of Web Inspector's functionality—in particular, you should try entering text directly at the console (a feature not available in JavaScript development, unfortunately). Can you execute `alertCompliment` from the console with a different message?

Moving to native platforms

Okay, so far so good, but to really appreciate PhoneGap we will need to create a native project based on the code that we have.

Time for action – You Are The Best for iPhone

We're going to build **You Are The Best** for iPhone now, but if you followed *Chapter 1, Installing PhoneGap*, you shouldn't see any surprises if you want to build to something else first:

1. Launch XCode and create a new PhoneGap-based project, **YouAreTheBest**, in a safe location.
2. Copy our **YouAreTheBest** assets into that project's `www` directory.

3. Hit **Build and Run** to launch the iPhone simulator:

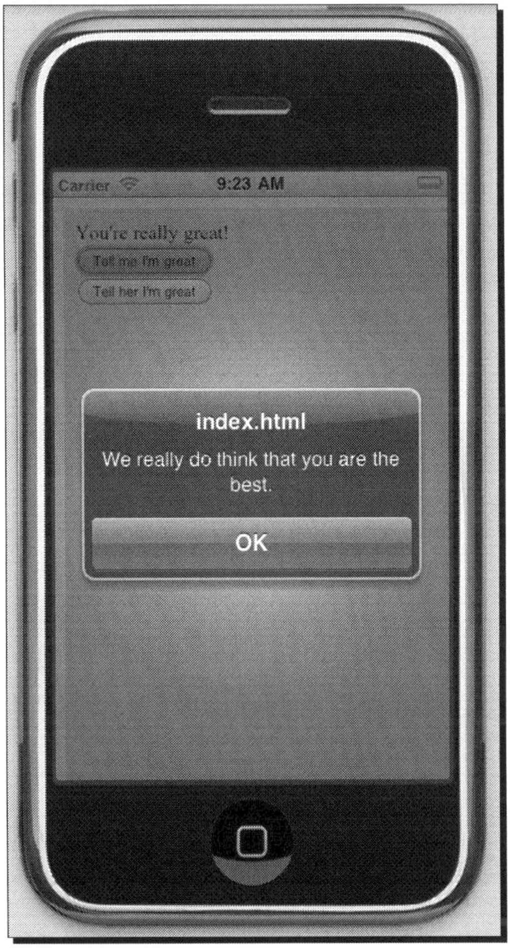

4. Not bad, but a couple of things stand out right away—the scale is way off, and the alert should be from `YouAreTheBest`, not from `index.html`.

5. Firstly, let's deal with the scale of the application. Mobile web views typically have a component called the **viewport** that handles things like scale. We'll touch on what the viewport actually does later on, but for now, add this tag to the `head` of our document:

```
<meta name="viewport" content="width=device-width, initial-
scale=1.0, user-scalable=no"></meta>
```

6. And we'll also modify our CSS to take advantage of the new dimensions:

```
div#main {
  background: #ccf;
  height: 82%;
  padding: 10%;
  width: 80%;
  -webkit-border-radius: 8px;
}
```

7. Back to the simulator, and things look a lot better already:

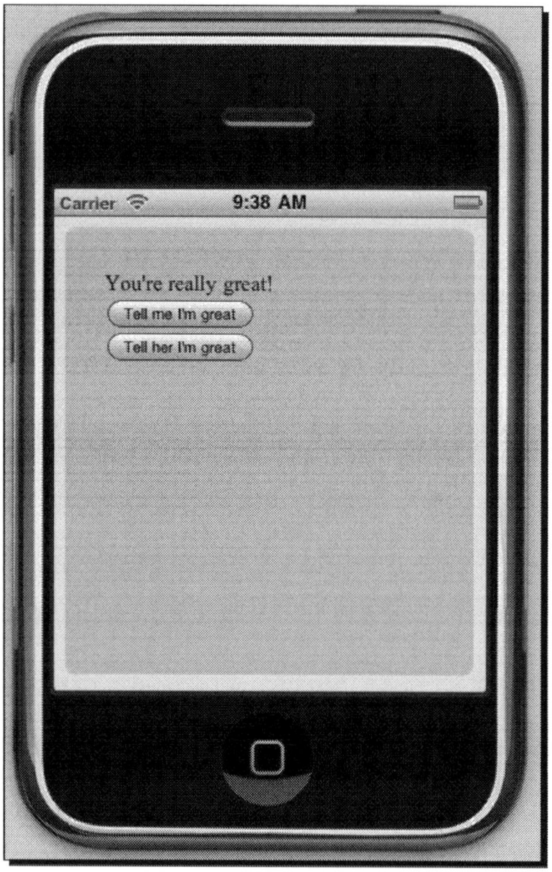

8. Now for that pesky alert box. This will be the first PhoneGap JavaScript call we make in **You Are The Best**. The first priority is to add the `phonegap.js` script to our document:

```
<script src="phonegap.js"></script>
```

9. If you list the contents of the `www` directory, you will notice that `phonegap.js` is already there—it gets automatically generated at build time.

10. We are next going to change our call to alert a PhoneGap specific call to `navigator.notification.alert`: we go from:

```
    alert("We really do think that " + greeting + " the best.")
```

    To the following:

```
    navigator.notification.alert("We really do think that " +
greeting + " the best.", function() {}, "You Are The Best")
```

11. `navigator.notification.alert` allows us to customize the alert box with a title and button name, so it appears less like a web notification and more like a native notice.

12. Please note that we're including an empty function as the second argument to the function; this is due to `navigator.notification.alert` taking a callback argument, for asynchronous execution of a callback code. This is distinct from the browser's `alert` call, which is blocking.

13. Because PhoneGap has to be initialized for that call, we're going to change the initialization event listener to listen for `deviceready` rather than `DOMContentLoaded`:

```
document.addEventListener("deviceready", function () {
// removed initial alert by now - it was getting annoying
        document.getElementById("likeMe").
addEventListener("click", alertCompliment, false);
  document.getElementById("likeHer").addEventListener("click",
alertCompliment, false);
}, false);
```

14. Now let's run that in the simulator and see if it all works:

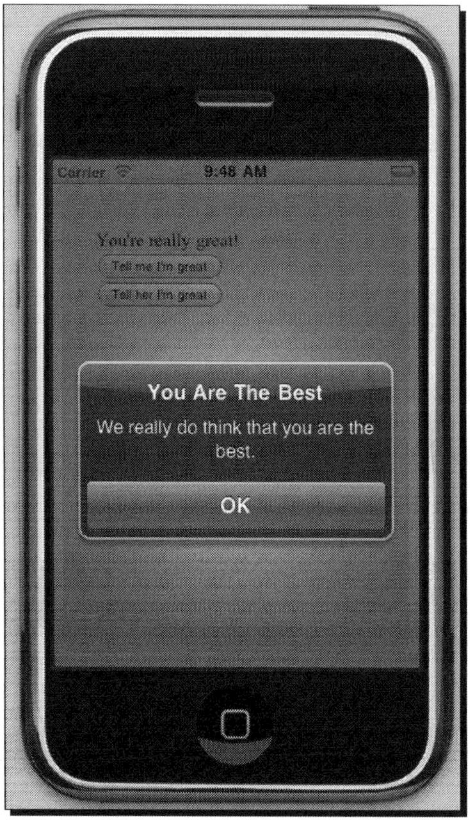

Hooray!

What just happened?

We took an application that was prototyped, developed, and debugged in a desktop browser and rapidly created a native mobile project from that application. While we could have stopped at what we created in Safari, we were able to make some small device-specific optimizations to enhance the look and feel of our application.

The key points were the changes we made to our desktop project:

<meta name="viewport">

The viewport is a little unintuitive at first, but very important to grasp for mobile web or PhoneGap development. Desktop browsers function in terms of windows: content is rendered to the size of the window, and the user navigates within the window. Mobile browsers—that is, post iPhone, usually WebKit-based mobile browsers—render their content to a window, but have an intermediary between the user and the window, which is the viewport.

The user can not only move the viewport to different parts of the window (as opposed to scrolling the window itself), but also zoom in and out to improve legibility, usability, and other abilities. Using this `meta` tag, web developers can also recommend how the browser's viewport should be rendered.

This is fantastic for navigating the web, but is a markedly different experience from using a native application, where you typically navigate only within a screen-sized fixed window. To get this kind of behavior, mobile web and PhoneGap developers typically use a `meta` tag with these properties:

```
<meta name="viewport" content="width=device-width, initial-scale=1.0,
user-scalable=no"></meta>
```

That's saying:

- Set the width of my page to the width of the device
- Don't do any initial scaling: I will design for the device width
- Don't let the user rescale

You may wish to edit that `meta` tag at a later point, but for now, I highly recommend sticking with it.

phonegap.js

The `phonegap.js` file provides JavaScript access to the native APIs exposed by the PhoneGap framework. We will cover PhoneGap throughout this book.

Okay, so there are a couple of points to emphasize about `phonegap.js`:

- By default, `phonegap.js` is automatically generated at build-time in the root of your www directory. If, for whatever reason, you want to edit this generated file, but not edit the source JavaScript files (if there's just a quick fix you want to make, say), I would recommend moving the file to a subdirectory of your www directory (`www/js/phonegap-modified.js`, or the like).

- The reason phonegap.js is generated like this is because it is platform-specific—you cannot use one phonegap.js file for iOS, Android, BlackBerry, and everything else. There has been some work toward getting a single unified JavaScript file, but we're still far away from that.

In some releases of PhoneGap, the JavaScript file will have the version of PhoneGap appended also: phonegap-0.9.5.js, phonegap-0.9.6.js, and so on. Naturally, you'll want to amend your script tags in these cases.

deviceready

One other significant change that we made from our desktop code was to subscribe to the deviceready event instead of DOMContentLoaded. deviceready is a PhoneGap specific custom event, fired on the document object, that informs the developer that PhoneGap is ready, and all of the native APIs can now be called.

deviceready always fires after DOMContentLoaded, usually less than a second. If you have any code that manipulates the DOM and makes PhoneGap calls, it can be safely called when deviceready fires. However, there is that delay to be aware of—if you need the code to execute as soon as possible and it doesn't use PhoneGap APIs, either load or DOMContentLoaded will be your best bet.

Have a go hero – Porting to other platforms

The obvious next step is to port the application over to some more platforms—Android, BlackBerry, HP webOS, and any other SDKs you can get your hands on. You should be able to look at the tutorials from the previous chapter to see how to set up these other platforms, and copy your files into those projects.

Please note that some releases of PhoneGap for BlackBerry WebWorks include the JavaScript file under www/javascripts/phonegap.js, rather than www/phonegap.js—it's a non-standard gotcha you'll want to be aware of.

What behavior surprises you when you port the application? Can you notice the differences between each platform's WebKit?

Can the same debugging techniques apply for debugging in a simulator—can you call console.log, and where does it write to? Is this call consistent between browsers? You should be aware of any quirks now, before you get involved in more complex application development. Make sure you check http://docs.phonegap.com, noting your PhoneGap version number, if anything surprising happens.

Summary

This chapter helped us learn a lot about the process of developing PhoneGap applications—how much time we need to spend in the native SDKs and emulators, and how much we can offload to a desktop browser. Even on iOS, arguably the most restrictive platform, we can do most of our work in Safari.

Specifically, we covered:

- Rapidly prototyping our application in a web browser
- Adding extra functionality and using the browser's debugging tools to ensure the new functionality works
- Bringing our application to a native platform, observing any new issues that arose, and fixing them in a sensible fashion

We also familiarized ourselves further with the structure of PhoneGap applications and the conventions that PhoneGap uses to maximize developer productivity—key skills that we'll need as we progress through the chapters.

Now that we've learned about how to build PhoneGap applications, we need to know what these applications are going to do. In the next chapter, we'll cover what, specifically, a PhoneGap application does that a mobile web application does not.

3

Mobile Web to Mobile Applications

We finally have a robust setup for developing cross-platform mobile applications using PhoneGap. Now what? How easily can you use your existing web development skills to build compelling mobile applications? Which parts of your workflow need to be changed?

This chapter will cover the changes that will turn your projects from great websites into great mobile experiences. In particular, we'll look at how to restructure a traditional browser-server site to take full use of a purely client-side environment, and fulfill most of the web server's roles locally. All of the techniques we use will be based on HTML5 standards and pure JavaScript, so they will be fully portable across all our supported platforms.

In this chapter, we will:

- Use HTML5 persistence APIs to store user data locally
- Template our views in pure JavaScript
- Communicate with network resources across domains

Implementing web server roles

In a traditional web application, the responsibilities for application behavior are divided between a remote web server and local web browser. The browser handles the user interface—ensuring that user actions have quick responses and behave as expected, while the server generates web pages and stores persistent data. This split responsibility allows the browser application to be light and responsive, while the server can handle more intensive tasks.

There's no technical reason for a PhoneGap application to depend entirely on a connection to a remote server in this way and, as we'll see later, the curious nature of PhoneGap's runtime environment means that remote server calls are unrestricted. However, if you have any firsthand experience with a smartphone, you will know that connectivity is far less reliable than on a full personal computer. A mobile device is used in plenty of contexts—in buses, airports, planes, and bathrooms—where a network connection can be a luxury.

With the JavaScript techniques we're going to explain, is a far less of a hindrance than you may think—in fact, it's a pleasure to write applications this way. We'll start with the most important piece: your user's data.

Time for action – Implementing LocalStorage

1. Let's create a new directory for this chapter—`sites_without_servers`—and create an empty `index.html` file.
2. We'll fill in our `index.html` file with the same kind of code we had in the previous chapter, but this time add a couple of form fields (and make the colors less garish):

```
<html>
  <head>
    <title>Sites Without Servers</title>
    <style>
          body {
            background: white;
          }
              div#main {
    background: #ccc;
          border: black 1px solid;
    height: 82%;
    padding: 10%;
    width: 80%;
    -webkit-border-radius: 8px;
  }
    </style>
    <meta name=»viewport» content=»width=device-width, initial-
     scale=1.0»></meta>
  </head>
  <body>
    <div id=»main»>
        <h1>List Some Food</h1>
        <form id=»foodForm»>
          <input type=»text» id=»foodName» placeholder=»A name of
           a food» />
          <button id=»submitFood»>Tell Us!</button><br>
```

```
        </form>
    </div>
  </body>
</html>
```

3. Now open this file in Safari and have a look:

4. You can write the name of a food in the field, press the button, and the page reloads. This is good place to improve upon.

5. Incidentally, you can see our use of the HTML5 `placeholder` attribute on the `input` element—giving a cue to our users, if the main header wasn't enough. It great in browsers that have support, and causes no harm to those that do not.

6. We're going to write some JavaScript to capture the submission of the food form, write the food's name out to the DOM, and clear the form field, instead of reloading the page. Everything here is standard browser JavaScript, and should be familiar from the last chapter:

```
<script>
    document.addEventListener(«DOMContentLoaded», function () {
      var foodList = document.getElementById(‹foodList');
      var foodField = document.getElementById(‹foodName');
```

```
            document.getElementById(«foodForm»).
addEventListener(«submit», function (evt) {
          evt.preventDefault();
          var newFood = foodField.value;
          var newFoodItem = document.createElement(‹li');

          newFoodItem.innerHTML = newFood;
          foodList.appendChild(newFoodItem);

          foodField.value = «»;
          return false;
       }, false);
    });
  </script>
```

We're also going to add a <ul> tag under the form, which we can append new foods to:

```
  <ul id=»foodList»>
  </ul>
```

7. Reload the browser, and enter a few foods to test this out.

8. So far, so good. Now reload the page again. We've lost our list of foods, our users are distraught, and we're now unemployed. Nobody is safe in this economy. Let's try to save our foods persistently.

9. We're going to use the **HTML5 LocalStorage API** here—we'll go into some of the other options shortly. We will add the API calls after the call to `appendChild`, as follows:

```
var foodKey = "food." + (window.localStorage.length + 1);
window.localStorage.setItem(foodKey, newFoodItem);
```

10. If you've gone to your browser to test things out, you probably haven't noticed any difference yet. Patience, dear reader! We need to revise the view code to display the saved foods. Since this code will also be about writing list items to the `foodList`, we'll abstract that part of the code into a separate function. Here's what the final JavaScript code looks like:

```
<script>
document.addEventListener(«DOMContentLoaded», function () {
    var foodList = document.getElementById(‹foodList');
    var foodField = document.getElementById(‹foodName');
    var l = window.localStorage.length;
    var i = 0;
    var storedFoodName;

    function addNewFoodItem(foodName) {
      var newFoodItem = document.createElement('li');
      newFoodItem.innerHTML = foodName;
      foodList.appendChild(newFoodItem);
    }

    for (i; i < l; i++) {
      storedFoodName = window.localStorage.key(i);
      if (storedFoodName.match(/^food[.]/))
      addNewFoodItem(window.localStorage.getItem(storedFoodName))
    }

        document.getElementById(«foodForm»).
addEventListener(«submit», function (evt) {
      evt.preventDefault();
      var newFood = foodField.value;
      var foodKey = «food.» + (window.localStorage.length + 1);

      addNewFoodItem(newFood)
      window.localStorage.setItem(foodKey, newFood);

      foodField.value = «»;
      return false;
    }, false);
  }, false);
</script>
```

11. Enter some foods, then reload the page to verify the food names are persisting now:

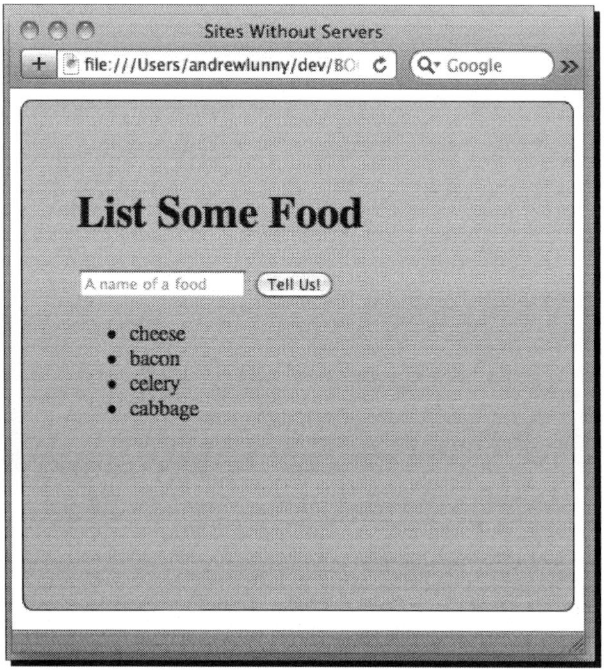

What just happened?

We got an introduction to the HTML5 LocalStorage API. The last couple of changes to the code, to display the stored data at load time, introduced a number of changes. Let's go through these one at a time:

```
var foodKey = "food." + (window.localStorage.length + 1);
```

This is the first time we access the `window.localStorage` object itself. It is not a JavaScript array—the object is an instance of the native `Storage` class which doesn't have constructor access to from inside the browser environment—but it has array-like features, such as a `length` property.

We generate a key to store each food by joining the prefix `food.` with `localStorage.length + 1`, arriving at a pseudo-unique string. It's not a sustainable algorithm, but it's suitable for these purposes, as no key should conflict with any others (our little app is append-only, for now).

There is no higher-level namespacing baked into `localStorage`—this ad-hoc mechanism (the `food.` prefix) is our best bet. All items stored from a domain are kept in the same blob—this is less of a concern with PhoneGap applications, where you know nobody else is writing or reading, but can be a concern in other environments, particularly shared hosting.

```
window.localStorage.setItem(foodKey, newFood);
```

LocalStorage is a key value store—the key is the variable we set in the previous line, and the value is the food entered by the user in our form. We like the cleanness of the LocalStorage API—there's no ambiguity about what this line is doing.

```
var l = window.localStorage.length;
for (i; i < l; i++) {
  storedFoodName = window.localStorage.key(i);
  if (storedFoodName.match(/^food[.]/))
    addNewFoodItem(window.localStorage.getItem(storedFoodName))
}
```

This is the logic to populate our list with the contents of `localStorage`. A few notes:

- Because `localStorage` has an array-like `length` property, we can iterate through its contents with a simple `for` loop. There is no more advanced querying mechanism, for better or worse.
- `window.localStorage.key` is a function to retrieve the key for the item located at a particular index. As we iterate through `localStorage`, we can use the key function to check every stored item.
- The `if` statement above checks if the key matches the pattern of the foods we have set—it should look like `food.1`, or `food.42`. It uses the regular expression `/^food[.]/` to match against the available keys—this regex checks if the string begins with `food.` If so, we use the `setItem` function to pull out the food's name and insert it into the DOM. Again, there's no namespacing in `localStorage`, so we need to check the contents of each key for assurance that is what we were looking for.

What can we deduce about LocalStorage from this example?

- **It's simple**: We were able to get the desired behavior up and running in fewer than 10 lines of code. The most complex thing we wrote was a loop to iterate through an ordered list.
- **It's good enough**: We can store the user's input, and utilize it whenever we want. We can also retrieve the input send back to a remote server, or do whatever we choose, if we have more powerful tools elsewhere.

- **It's synchronous, or blocking**: Our keys and items are stored in memory, so we have immediate access to them in most cases. This is nice for small use cases like the above, but may incur problems with more intensive use. For example, if we're retrieving very large objects from LocalStorage, the single JavaScript thread will be occupied doing this, and no other code (for example, responding to user actions) can execute.
- **It's dumb**: There's no querying language, no indexing, no storing of rich objects. Anything more powerful has to be implemented at the application layer.

From a cross-platform perspective, LocalStorage is the best supported option, and is highly recommended unless your needs overwhelm it.

Other storage options

How long do you have? Let's skip over the oldest persistent storage hacks—cookies and the venerable `window.name` hack—and talk about HTML5.

Web SQL

LocalStorage was initially paired with an interface referred to as **Web SQL**, which is currently well-implemented on first-class mobile devices (iOS, Android, and BlackBerry OS 6.0). Web SQL is a thin, asynchronous wrapper around an **SQLite** database—SQLite, if you're not familiar, is an SQL storage interface to a flat file, rather than a database server. Web SQL works well if your remote application server uses an SQL database, and you want to mirror the structure of your data between platforms. And if you're familiar with SQL, it offers a useful interface for storing structured data.

However, Web SQL was never too popular. SQL data-types map poorly to JavaScript objects, the long strings of SQL literals are particularly error-prone, and the interface was frustratingly tied to a single version of a single SQL implementation. For these reasons, and probably others, there was never a full Web SQL spec and it was deprecated in November 2010.

Indexed DB

In its place, a competing spec called **Indexed DB** has been promoted. Indexed DB is a schema-free approach for storing JavaScript objects (as opposed to strings, or SQL data types) persistently, based on an indexed property on the object itself. Stores of objects can be iterated over easily, and indexed on multiple fields. It's a flexible enough option that we expect very powerful database interfaces to be built right on top of the native interface.

The problem? Indexed DB doesn't really exist yet (at the time of writing). The spec itself has not been finalized; there are incompatible, vendor-prefixed partial implementations in the latest releases of Firefox and Google Chrome, and no mobile platform ships with even an experimental implementation. So for now, it's a non-starter.

So one deprecated option, and one unimplemented option. LocalStorage isn't looking so bad now, right?

This covers the situation on the first-class mobile platforms—on older platforms that do not support any of the HTML5 spec, you will need to fall back to either cookies or to a native solution. For example, on the BlackBerry OS 5.0 and below, PhoneGap includes an interface to the native key-value store that you can access from JavaScript; there is a similar interface for PhoneGap's `Symbian.wrt` release. And if all else fails, you can use PhoneGap's `File` API to read and write directly to the file system—but this is likely to be the slowest and most expensive option.

A word of recommendation should go out to **Lawnchair**, a storage library written by Brian Leroux from the PhoneGap team and compatible with all of the supported PhoneGap platforms. Lawnchair gives a unified key-value store API to all of the storage options listed above—if you find yourself writing code for multiple different local storage backends, give Lawnchair a shot. You can get the code at `http://github.com/brianleroux/lawnchair`.

Have a go hero: Exploring LocalStorage

Look at the instructions for building native applications from the first two chapters, and get this application running on whatever device you have at hand—an iPhone, an Android, a BlackBerry device, or whatever. Does the basic functionality still work—once you submit foods through the form, are they reflected on screen?

Try opening and re-launching the app, manually ending the application if necessary. Are the values persisting? What happens if you uninstall the application? What about if you reset the device?

You should have a firm grasp of the `localStorage` API at this point—let's implement some other web server roles.

View templating

The next item on our laundry list of web server roles to port over is, unlike persistent storage, not implemented as an HTML5 API. View templating is the kind of software task that traditionally lends itself to **bike-shedding**—that is, an awful lot of debate over very small differences. It's easy to implement, there are innumerable functioning alternatives, and it's necessary for every sizeable project. A non-exhaustive list of available libraries can be found on Mozilla's Developer Network, at `https://developer.mozilla.org/en/JavaScript_templates`.

We are going to use the **Mustache** templating language, implemented in JavaScript by Jan Lehnardt. The code is available at `http://github.com/janl/mustache.js`. There's a lot to like about Mustache—it cleanly separates the templates themselves from any view logic, and is available in server-side and client environments—but if there's a JavaScript templating approach you prefer, feel free to use that one.

`Mustache.js` is also the first external library we'll be using, so best of luck!

Time for action – Food detail view

1. First things first, let's get a local copy of `mustache.js`. We could clone the entire Git repository, but we only need a single JavaScript file. We can grab that one directly from `https://github.com/janl/mustache.js/raw/master/mustache.js`. Download that file and save it in your `sites_without_servers` directory.

2. Next, include that file in your `index.html`—add this `<script>` tag just before your closing `</body>` tag:

```
<script src="mustache.js"></script>
```

3. Now let's write our first Mustache template. Create a new file called `food_detail.mustache` and fill it with the following code:

```
  <h2>{{ foodName }}</h2>
  <p>{{ foodName }} is my favorite food. I particularly enjoy it
in the {{ timeOfDay }}.</p>
```

4. This file is known as our **template**, by Mustache's (straightforward) terminology. We need to pair our template with a **view** in JavaScript that populates its contents. So let's write our view as follows (you can add this code after the opening `<script>` tag in `index.html`):

```
var aFoodDetail = {
  foodName: «cereal»,
  timeOfDay: function () {
   return this.foodName.length > 7 ? «evening» : «morning»;
  }
}
```

5. You probably don't decide your personal eating preferences based on the length of the food's name, but it illustrates one of the nice features of Mustache—we can define fields in our view either as JavaScript objects, or as functions to be executed when we render our template.

6. Now we need to combine our view and our template to get our output. We've chosen to store our template in a separate file, so we need to get the contents of that file first, and then render the view within those contents. Ensure this code is in the top-level scope of a `<script>` tag, so that we can easily call the functions from the Web Inspector:

```
var storedTemplate = null;

function renderOurTemplate(view) {
  function doRender(template, view) {
    console.log(«rendering now»)
    console.log(Mustache.to_html(template, view))
  }

  if (storedTemplate) {
    console.log(«template is stored - we can render
    immediately»)
    doRender(storedTemplate, view);
  } else {
    console.log(«template isn't stored - need to request it»);
    var req = new XMLHttpRequest();
    req.onreadystatechange = function () {
      if (this.readyState == 4) {
        if (this.status == 200 || this.status == 0) {
          storedTemplate = this.responseText;
          doRender(storedTemplate, view);
        } else {
          console.log(«something went wrong»);
        }
      }
    }
    req.open(«GET», «food_detail.mustache», true);
    req.send();
  }
}
```

7. There's a lot going on there that we'll cover in detail, but first let's verify that it works.

8. Reload our `index.html` file in Safari, and open the Web Inspector. Enter this command:

```
$ renderOurTemplate(aFoodDetail)
```

9. And check the output—there should be the message **template isn't stored—need to request it**, and then the rendered Mustache template, logged to the console. Try running the function again:

```
$ renderOurTemplate(aFoodDetail)
```

10. And this time ensure that **template is stored—we can render immediately** is printed before the rendered code.

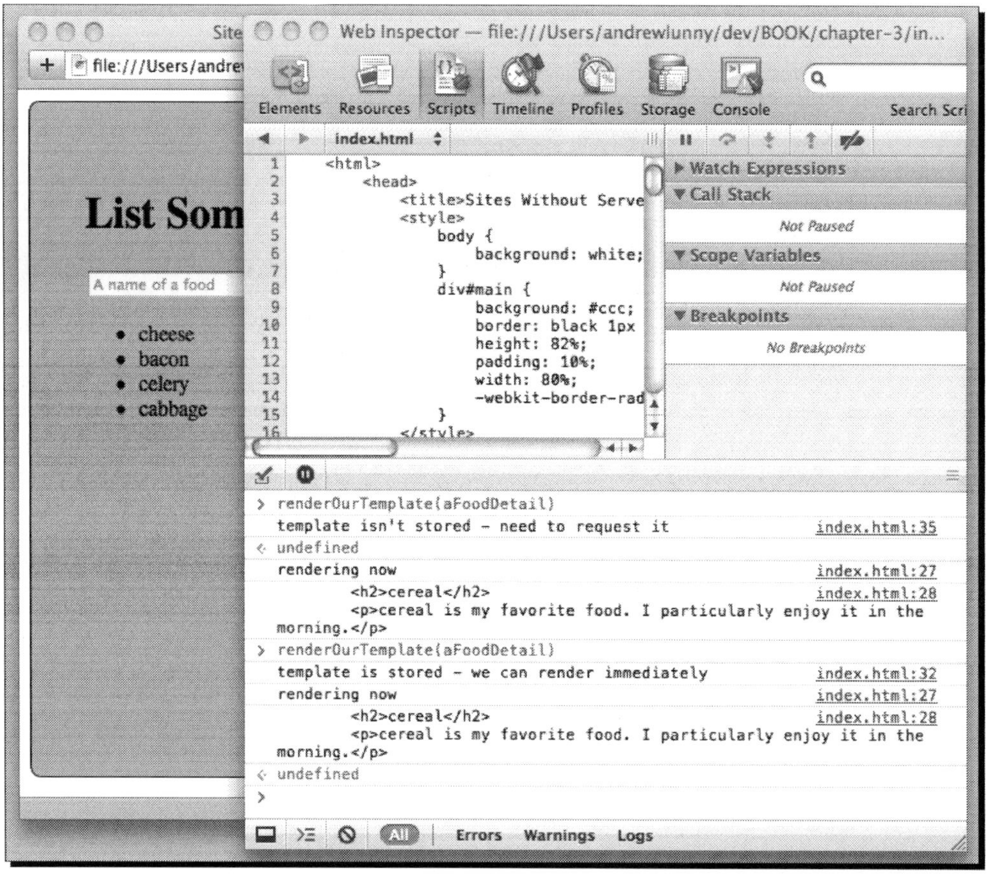

11. Now the next step is to insert the rendered markup into our document. First, find the `<ul>` element that was there from our previous example, and insert an empty `<div>` above it:

```
<div id="foodDescription">
</div>
```

11. We next want to modify our `renderOurTemplate` function to return back the rendered HTML. Since we may have to make a request to get our template, we get to do this call asynchronously, and only need to change the first few lines of `renderOurTemplate` to achieve this:

```
function renderOurTemplate(view, callback) {
  function doRender(template, view) {
    console.log(«rendering now»)
    callback(Mustache.to_html(template, view))
  }
...
```

12. Now we can modify our `addNewFoodItem` function, to write the food detail to our page as well as entering each new food onto the list. Here's what the function will now look like—the changes are highlighted:

```
function addNewFoodItem(foodName) {
  var newFoodItem = document.createElement(‹li');
  newFoodItem.innerHTML = foodName;
  foodList.appendChild(newFoodItem);

  aFoodDetail.foodName = foodName;
  renderOurTemplate(aFoodDetail, function (markup) {
    document.getElementById("foodDescription").innerHTML =
markup;
  });
}
```

13. Note that we're changing the `foodName` property of our `aFoodDetail` object to reflect the user input, and then writing the result to the DOM. Let's reload Safari and... well, it renders right away, since it's hooked into the same `addFoodItem` function as the list of foods. That's convenient!

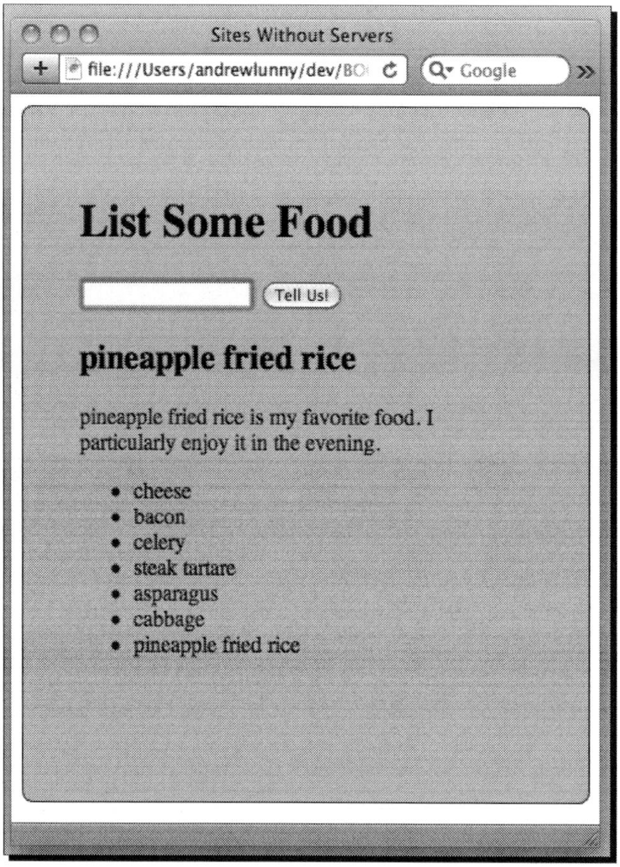

And now our content is beautifully rendered!

What just happened?

We've done a few things just now:

- Created an HTML template to display our application data
- Used Mustache to dynamically populate that template at runtime
- Integrated Mustache with our existing code to render user data according to our template.

A very simple technique, but an enormously useful one.

View templating isn't difficult—all we've done here is taken an HTML fragment with some placeholder content, replaced that placeholder content with live content from our application, and rendered the populated HTML into our page. With libraries like Mustache, once we have our content and our template in the right format, it's just a single function call to get our output data ready to go.

Also, to re-emphasize a point made prior to the tutorial, the basic premise applies to any templating library you'll use, or even one you may write. Mustache is just one example of a library that takes care of the grunt work.

The main tricky point in the above example was getting the template into JavaScript from the filesystem. There's a lot to love about JavaScript, but one of the major things to hate about it is the lack of support for multiline strings, or Heredocs as they're often known in other languages. If our two line template was a JavaScript variable, it would look like this:

```
var foodDetail = "<h2>{{ foodname }}</h2>" +
  "<p>{{ foodname }} is my favorite food. i particularly enjoy it in
the {{ timeofday }}.</p>";
```

This is fairly ugly with a two line file, but hugely ugly, brittle, and unmaintainable once you go beyond 10 or 15 lines. And since the Strings can't be unescaped either, all double quotes have to be manually escaped (the same would apply to single-quotes, if those surround your string literals).

Instead, we put our HTML fragment in a separate file and use an `XMLHttpRequest` to load it at run time. This makes our view code easier to read and modify, and only has a minimal performance effect on the application as a whole. The actual loading mechanism is a little opaque; converting the JavaScript code to pseudo-code makes it easier to see what's happening:

```
if (the template is in memory)
  render the view on the template
else
  request the view, and then
  render the view on the template
```

The actual code is more complicated since the rendering is done asynchronously, and uses the fairly verbose `XMLHttpRequest` API, but the basic idea is to implement a very simple caching mechanism when loading our template into memory. We're requesting a local resource, so the actual AJAX request is very inexpensive, and it only needs to be performed once for each template in your application.

The rest is much the same as the previous chapter—taking data and inserting it into the DOM, using the standard browser APIs to display interactions with the user. If you're an experienced web developer coming to PhoneGap to reach mobile devices, all of the code should be comfortable and familiar.

Pop Quiz – Templating with Mustache

1. What advantages does Mustache have over simple string interpolation?
    a. Very few: you could do all of this by concatenating string literals.
    b. It gives an optional compatibility layer with your server templates.
    c. It handles the standard templating features with speed and consistency.
2. When is the best time to render a template for display?
    a. At load time—insert it into the DOM, and then hide it with CSS.
    b. When there are some spare CPU cycles—if the user is idle, do some rendering.
    c. Just in time—render at once when the user requests a view.

Our list of foods, under the `foodList` element, uses `document.createElement` to insert new elements directly into the DOM. What are the pros and cons of using this approach, as opposed to a template file and a Mustache view?

Accessing remote resources

The last of the three major web server roles we will look at will be accessing remote resources.

Our section on view templating gave an introduction, in case you were unfamiliar, to the **XMLHttpRequest (XHR)** object in JavaScript. As the foundation of Ajax, the world's greatest pre-HTML5 JavaScript buzzword, XHRs allow us to retrieve arbitrary resources from HTTP servers without doing a full page reload. Although newer innovations, most notably the WebSockets spec, have stolen some of the XHR thunder, it's still an invaluable tool for the JavaScript developer.

Cross-origin policy

If XHRs are so powerful, why do web servers still handle most external network requests? The shortest answer is that XHRs are bound by the browser environment's **cross-origin policy**: a resource (an HTML page) can only access other resources from the same origin (the same URL scheme, the same hostname, and the same port). If my page is served from `http://alunny.ca/foo.html`, I can't make an XHR request to `http://search.twitter.com` (or `https://alunny.ca`, for that matter).

Interestingly, there are changes in the HTML5 spec to allow for cross-origin resource sharing– these require setting HTTP headers on the server that gets accessed, and are outside the scope of PhoneGap application development. There is a similar spec for cross-domain messaging, between frames from separate domains in the same browser page; but again, this is not applicable for PhoneGap applications.

The cross-origin policy, however, is not a concern for PhoneGap applications, where our pages are not served from a web server. Luckily, the policy is not enforced on the `file://` protocol—that is, a page served from `file:///Users/andrewlunny/foo.html` can make an XHR request to `http://search.twitter.com`.

This is broadly true for all of the mobile platforms that PhoneGap supports—the one notable exception is the native BlackBerry port (which exists for BlackBerry OS 4.6 compatibility, chiefly), and PhoneGap implements a `Network.xhr` function to do the equivalent requests. You will have to use both functions if you want to target BlackBerry OS 4.6, along with more powerful platforms—your best bet is to write a simple wrapper function that chooses the appropriate method based on the current execution environment.

One final sidenote: the cross-origin restrictions may end up in place on the `file://` protocol after all, as they're beginning to do on desktop browsers. Apple's Safari browser, which we have been using for development, has some spotty issues with version 5 (version 4 allowed cross-origin requests, much as PhoneGap platforms do); it has been working with the demo I've prepared, so we'll stick with it for now. For PhoneGap, the dev team has committed to patch each platform if these restrictions appear; we'll use a workaround for doing the same on a desktop browser shortly.

Time for action – Talking about food

Back to our little app—we've done a great job of listing the foods that we like, but what do other people think of our favorite foods? Wouldn't it be great to know what everyone in the world is saying about the food we love?

Luckily we have an amazing new invention called **Twitter** to help us. Twitter is a popular website that allows celebrities to demonstrate their ignorance to thousands and regular folk to share their insights with coworkers. Although Twitter's timeline API requires authentication to access (and is a bit frustrating to get working with PhoneGap), it's search API is wide open, and is a great one to play around with.

1. Do a sanity check API call—we want to ensure the Twitter search API is online, available, and returning the right kind of results. Open a Terminal window, and use the **curl** tool to query Twitter's search API; if you're on Windows, you will probably want to use a shell like **Cygwin** or **Git Bash** to get the best results.

```
$ curl http://search.twitter.com/search.json?q=bacon
```

2. We should get a large JSON response back, which will look something like this:

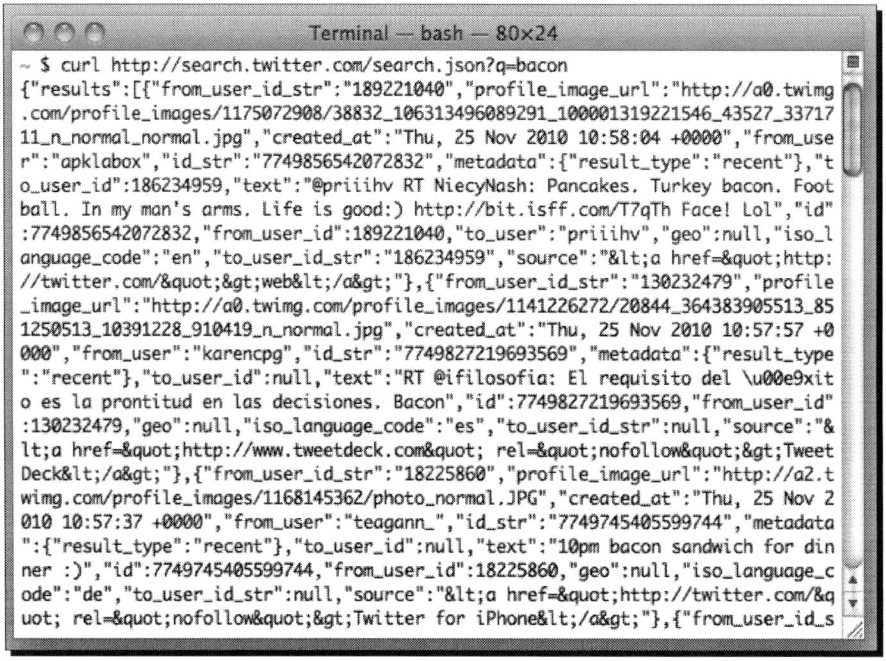

Hey, bacon's a very popular food.

3. Now let's bring that API call into our PhoneGap application. Firstly, we're going to simplify the logic to call `XMLHttpRequest.send` from our application, with a new `getXHR` function:

```
// performs a get request for url
// passes the response text to callback
function getXHR(url, callback) {
  var req = new XMLHttpRequest();
  req.onreadystatechange = function () {
    if (this.readyState == 4) {
      if (this.status == 200 || this.status == 0) {
        callback(this.responseText);
      } else {
        console.log(‹something went wrong');
      }
    }
  }
```

```
  req.open(<GET', url, true);
  req.send();
}
```

4. As usual, ensure that function is in the global scope, so we can access it easily from Web Inspector.

5. Do just that—open our page in Safari, launch Web Inspector, and enter the following:

```
$ function log(x) { console.log(x) }
$ getXHR('http://search.twitter.com/search.json?q=bacon', log)
```

6. If you're running into cross-domain restrictions—errors such as **Failed to Load Resource**—you can use Twitter's JSONP interface instead. JSONP allows you to access JSON resources through an external `script` tag, rather than with an `XmlHttpRequest`. First, we'll define a function `getJSONP`:

```
// loads an external url through a <script> tag
// avoids cross-domain restrictions
function getJSONP(url, callback) {
  var s = document.createElement(«script»),
      path = url + «&callback=jsonpCallback»;
  window.jsonpCallback = callback;

  s.src = path;
  document.body.appendChild(s);
}
```

7. Then call `getJSONP` instead of `getXHR`:

```
$ getJSONP('http://search.twitter.com/search.json?q=bacon', log)
```

8. The output should look like this (of course, we're getting live tweets, so the exact content will be different):

9. That's all well and good, but, if we're using `getXHR`, it's just returning a string—a JSON string, which we can work with, but a string nonetheless. If we're using `getJSONP`, we parse the object when the new `script` element is created.

10. Instead of logging the data we get back and doing nothing else, let's write a smarter callback function that actually parses the data we receive, if that's necessary:

```
function parseAndLog(JSONstring) {
   var JSONobj = JSONstring
   if (typeof JSONstring == "string")
JSONobj = JSON.parse(JSONstring);
  console.log(JSONobj);
}
```

11. Please note that for this to work, our browser will need to have the JSON object available—all modern desktop browsers will have it, but mobile devices will not. In these cases, you can use Douglas Crockford `json2.js` library—available at `http://www.json.org/js.html`—to patch in this functionality.

12. Cool, let's try that in Web Inspector:

```
$ getXHR('http://search.twitter.com/search.json?q=bacon',
parseAndLog)
```

13. And examine the results:

14. We can see we're now logging a JavaScript object, rather than a long string. This is great; we can easily manipulate and render this structure from here.

15. How can we integrate this easily into our existing app? Here's what we'll do: we'll add Twitter integration to the user experience that we've already created, like so:

    - When our user clicks, or touches, the name of a food.
    - We query Twitter for the latest tweet mentioning that food.
    - Then we display that Tweet in an annoying `alert` to the user.

    Works for me!

16. Let's get going—first, we're going to want to restructure our callback function to get the latest tweet about a food. Since the Twitter search API returns an array of results, with the newest result first, this is an easy function to write:

```
function getLatestResult(JSONstring) {
  var twitterPayload = JSON.parse(JSONstring);
  var latestResult = twitterPayload.results[0];

  return latestResult;
}
```

    See, no trouble at all.

17. Next, let's write the event handler for the clicks on `<li>` elements. We're going to use a technique called **event delegation** to catch all of these events—we listen on the `document` element, and then, if the target of the event matches one of our list items, perform our magic:

```
document.addEventListener("click", function (evt) {
  if (evt.target.tagName == «LI») {
    var foodSubject = evt.target.innerHTML;
    var foodSearch = encodeURIComponent(foodSubject);
    var twitterUrl = «http://search.twitter.com/search.json?q=» +
    foodSearch;

    getXHR(twitterUrl, function (response) {
      var latestTweet = getLatestResult(response);
      var msg = «Latest Tweet about « + foodSubject + « from « +
        latestTweet.from_user + «: « + latestTweet.text;

      alert(msg)
    })
  }
}, false);
```

    It's a little complex, but hopefully you can follow what's happening.

18. Now let's load it in Safari and give things a couple of clicks. If you're like me, you'll click on the names of the different foods a few different times, until you find a tweet that's innocuous enough to take a screenshot of and insert into your book on PhoneGap. Here we are:

And we now have a social app! Quick, somebody blog about this!

What just happened?

There were three separate, but overlapping, topics we covered in this section:

- Accessing remote resources through an `XMLHttpRequest`
- Parsing remote data as JSON, using the global `JSON` object
- Using event delegation to generate dynamic queries to a remote resource

Let's discuss each of these steps in more detail.

Accessing remote resources

We looked at a deliberately simple example in Twitter's search API—you can make an HTTP request from anywhere to that API and get some live data back. There are other considerations we need to have in most cases, however:

- **Authentication**: Purely public APIs are the exception, and most interesting work you'll do with remote APIs will have to be authenticated. Does the remote server use basic HTTP authentication, OAuth, OAuth 2, or its own custom authentication setup? Is SSL required for all authenticated requests? Do cookies need to be passed back and forth on every request?
- **Ownership**: Does the remote server belong to your organization, to an organization you're working with, or to a large Internet company (a la Twitter or Facebook)? Larger APIs tend to excel in terms of reliability and stability, while a remote server you have control over, or access to, clearly offers the most amount of flexibility. In between these extremes, there's a lot of scope for opaque and volatile APIs to beware.
- **Reliability**: This covers both uptime (how many of your requests will fail to complete) and also rate-limiting. Rate-limiting is something to be aware of in terms of the particular API calls you're making—the Twitter API, for instance, limits the amount of requests for user icon images far more aggressively than, say, requests for the public searches.
- **The API itself**: Is it a RESTful API or an increasingly rare SOAP one? Does the API provide XML, JSON, both, or other representations of data? Is there an easy mechanism to go from one call (get me the latest tweets about bacon) to a related call (get me the latest tweets from the guy who was talking about bacon)?

The Twitter search API, as an outstandingly simple example, is fairly portable—we should be able to execute basically the same code on any platform that supports cross-origin XHRs. With more complex access rules—particularly those based on authentication or other custom HTTP headers—older platforms will start to struggle. The PhoneGap `Symbian.wrt` port, for instance, requires the user to manually enter authentication details for authenticated requests.

The *Time for action* tutorial began with the use of `curl` on the command line and, when APIs expose their interfaces cleanly enough, a tool like `curl` is the best way to get started with remote APIs, using trial and error to explore the responses you will receive, and how you will need to format your requests.

Finally, in all this talk about remote APIs, it's worth emphasizing that this is all just HTTP—your request doesn't know, and shouldn't care, much about what's going on in the remote server. We could just as easily rewrite our above example to query a static file host, and parse a dump of JSON data that was stored as a flat file. Typically, however, larger APIs require at least some understanding of how they are implemented, and how that limits your interactions as an API consumer.

Parsing remote data

We touched on it just now, but one of the good things about the Twitter Search API is that it returns results in **JavaScript Object Notation** format, better known as **JSON** (pronounced "Jason"). JSON, as discovered and coded by JavaScript wizard Douglas Crockford, is a subset of JavaScript's object literals used to transmit strings, numbers, Boolean values, arrays (or ordered lists), hash tables (or dictionaries), and the null value. There are some important differences between JSON and JS object literals that JSON producers should be aware of, but as JSON consumers, we can ignore those for now.

JSON's elegant, lightweight syntax makes it a human-readable alternative to XML for data transfer, and reduces the amount of network bandwidth required to transmit data. There are JSON parsers and serializers implemented in every popular programming language. Although we still use the `XMLHttpRequest` to communicate with remote services, the majority of web APIs nowadays offer JSON data, alongside XML or as the sole option (such as the Facebook Graph API).

ECMAScript 5, the latest specification for the JavaScript language, defines a global `JSON` object that we used in the code above. The JSON object has two functions we care about:

- `JSON.stringify`: Takes a JavaScript object (including a primitive value) and converts it to a JSON string
- `JSON.parse`: Takes a JSON string and converts it to a JavaScript object

Using the native JSON object is the safest and most predictable way to use JSON data in your application, and the latest releases of first-class mobile platforms expose this object to developers.

Of course, with mobile development, it's never that easy. As you might expect, older BlackBerry devices and Symbian devices do not support the JSON object natively, but nor do iOS devices below OS 3.2 (all first generation iPhones and iPod Touches, for instance). A JavaScript implementation, JSON-js, is available at `http://github.com/douglascrockford/JSON-js`, and creates a JSON parser and a serializer you can use; you'll want to include the file `json2.js` from that repository, and add an appropriate `<script>` tag to your `index.html` file, to take advantage. Some PhoneGap platforms, including the BlackBerry port, include `json2.js`, so you may already have it at hand.

However, since JSON is just a subset of the JavaScript object literal notation, you can use the native `eval` function in JavaScript to convert a JSON string to JavaScript. This is the easiest (that is, most readily available) way to parse JSON in JavaScript, but is less than optimal for a couple of reasons:

- `eval` executes the string that is passed as JavaScript—so the call `eval("alert('foo')")`, for instance, will display that alert box, rather than throwing a parse error. If you don't trust the producer of the JSON you're using, don't use `eval`.
- `eval` takes strings that represent JavaScript expressions: in the case of object literals, the braces are parsed as block delimiters, rather than delimiting an object definition. This code, for example will throw a **SyntaxError**, since `"a":12` is not a valid JavaScript expression:

  ```
  var foo = '{"a":12}';
  var fooObj = eval(foo);
  ```

  We need to use parentheses to indicate that the string represents an object literal:

  ```
  var foo = '{"a":12}';
  var fooObj = eval(‹(‹ + foo + ‹)');
  ```

Issues like these, along with more obscure scoping problems, are why `eval` has such a bad reputation in JavaScript circles, and JSON-js is the best fallback when a native JSON implementation is unavailable. But if you're really tight on resources, `eval` is a good fallback to have.

Event delegation

The third approach to touch on slightly is how we bound the `click` events to each of our list items. Here's that code, simplified a little:

```
document.addEventListener("click", function (evt) {
  if (evt.target.tagName == "LI") {
    var foodSubject = evt.target.innerHTML;
    var foodSearch = encodeURIComponent(foodSubject);
    var twitterUrl = "http://search.twitter.com/search.json?q=" +
    foodSearch;

    // query the server and process the tweets
  }
}, false);
```

The first two arguments to addEventListener should be familiar; the third specifies whether the event should be captured (fires first on the document, then goes down to each contained target element) or bubble (fires first on the target element, then bubbles up to each containing element, up to the document). Event bubbling, which we specify explicitly by passing false, is the default, and is usually what we want.

Let's contrast that with how we bound the submit event of our form on the page (again, the code is simplified):

```
document.getElementById("foodForm").addEventListener("submit",
function (evt) {
  evt.preventDefault();
  var newFood = foodField.value;

  //process newFood
  return false;
});
```

In both cases, we use the addEventListener function to listen for a particular DOM event. For the submit event, we bind addEventListener to our foodForm element (the food form). When the submit event is fired on foodForm, we prevent the default action (by calling evt.preventDefault()) of submitting the page back to the server, do what we need to do, and then return false (so the event doesn't propagate).

For the click event, we listen on the document object, rather than on a particular DOM element on the page. When any user clicks on the document, our event handler function is called.

The evt object that's passed to this function, like any JavaScript Event object, includes a target field, which refers to the particular DOM element the event was fired on. Our logic is intended to execute for every <li> element in the document, so we check whether the target element matches our criteria and, if so, we perform our magic. This is known as **event delegation**: listening on a higher level for events, and delegating the event handler based on the event's target.

This approach offers a few benefits:

- We only need to set the handler once, for every element that we want to bind to. This avoids duplication of code, and any mistakes we might make if bind the wrong elements.
- It isn't dependent on the DOMContentLoaded event, since we bind it to the document object, we don't need our target elements to be initialized when we bind the event. We can set up our event delegation whenever we want.

- It binds the same events to new elements that are added to the DOM at runtime—it's the same pattern that's used, for instance, in the jQuery library's `$.live` and `$.delegate` functions. Again, this simplifies the amount of manual management we need to do on each element.

You can use event delegation for all events that occur on your application, but that may be a little much to manage at this stage—a good rule of thumb is to use a delegation pattern for similar events that need to fire on multiple, similar DOM elements, and directly bind event handlers that only relate to a single DOM element.

Sleight: The PhoneGap development server

One last thing to note on our use of remote resources: if you're having any trouble getting external `XMLHttpRequests` to execute, you may want to look at **Sleight**, a tiny web server we've developed to allow for external network requests from a single domain.

Sleight uses the **Node-JS** JavaScript platform to run its server, you'll want to install Node from `http://nodejs.org`, and also `npm`, the Node package manager, from `http://npmjs.org`. Once these tools are installed, you can install Sleight through `npm` as follows:

```
$ npm install -g sleight
```

Sleight serves static files from the directory you start it in, and also proxies remote requests to a remote server that you specify. For our app, we would want `http://search.twitter.com` to be the remote server, and we would start Sleight like so:

```
$ cd sites_without_servers
$ sleight port=4000 target=search.twitter.com
```

We could then navigate to `http://localhost:4000` to see our `index.html` file, or view it from our mobile browser (assuming our mobile device is on the same network as our development system).

In our JavaScript file, we could then manage our XHRs as follows:

```
var isOnFileProtocol = window.location.href.match(/^file/)
var domain = isOnFileProtocol ? "http://search.twitter.com/" : ""

// when we have a request to make
runXhr(domain + "search.json?q=bacon")
```

If we're running from the `file://` protocol (in a PhoneGap environment), we can do a cross-origin request; otherwise, we use Sleight to proxy the resource through a local web server.

You can find Sleight at `http://github.com/alunny/sleight`; it's not much code, but it can be a very useful development tool.

Have a go hero: Becoming more efficient

Although we looked at our three server replacements discretely, and they can be used entirely independently, there are a number of useful convergences between the three. In our view templating section, for instance, we cached our views in a JavaScript variable in memory, but we could just as easily have persisted the template to `localStorage`.

Try rewriting our Food List application with all three approaches in mind from the start. Here are some features you can implement:

- **Caching on remote requests**: If there's a network connection (remember the `navigator.Network.isReachable` PhoneGap call) get the latest tweet, otherwise pull one from `localStorage`.
- Access your view templates from a remote server, try putting your Mustache templates on a public server, and editing them while running the app. Is it possible to change the displayed template while the app is running? Would you want to keep the cached templates, even if the remote ones change?
- How would we go about backing up `localStorage`? Imagine our app has a database-backed web server we can access when we have a network connection—how would we try to sync the lists of food between both sources? How could we resolve conflicts in content? The answer is likely to be heavily dependent on the server implementation, but it's worth considering what the best approaches may be.

Summary

This chapter took three of the most important pieces of a server-based web application and moved them over to a robust thick client application, which we can easily deploy to a native mobile application with PhoneGap. Those three roles are:

- **Storing persistent data**: In our example, based on user input, but as we saw, we can easily modify our storage approach to include remote data, application views, or anything else that's textual.
- **View templating**: Using a library like Mustache, we can present dynamic and responsive user interfaces that can be updated easily.
- **Accessing remote resources**: With the power of cross-origin XHRs, we can create complex mash-up style apps that use remote APIs from anywhere on the web to enhance our users' experience.
- **Accessing remote resources from desktop browsers**: Either using JSONP where available, or the Sleight development server as a fallback.

Cumulatively, these three techniques allow us to maintain a stateful and responsive application that can take advantage of a remote server for data, but allow full user interaction within the context of a mobile application, be that online or offline.

In the next chapter, we'll combine these skills with some robust code management techniques to ensure our code portably and flexibly runs across all the devices we target, and can be easily modified to take full advantage of each platform's unique strengths.

4

Managing a Cross-Platform Codebase

PhoneGap provides some obvious pitfalls for developers—in particular, it's easy to write a cross-platform application that aims for the lowest common denominator, providing a mediocre experience that's identical on each device. Our goal, instead, should be to use the power and flexibility of PhoneGap to tailor the same codebase to take full advantage of each platform.

In this chapter, we'll see some tactics for achieving this goal, including:

- Using standard web development best practices, like feature detection and CSS media queries, to modify our application's behavior at run time
- Sniffing the platform at run time to conditionally execute or load JavaScript code
- Pre-processing our application code to target different platforms efficiently

Inherent differences between platforms

PhoneGap, we should never fail to stress, is not a write once, run anywhere solution for mobile development. Like web development in general, using PhoneGap allows you to easily get something up and running just about everywhere, but requires attention and diligence to ensure optimal performance and functionality for each particular environment.

There are two kinds of differences between platforms: ones we can detect at run-time and ones that require external knowledge. For example, we can feature-detect support for CSS3 transforms, but there's no way to detect if those transforms are hardware-accelerated or purely software—that data isn't exposed to application code. At the time of writing, iOS *does* accelerate those transitions (giving a much smoother experience), while all other platforms run the animations through software, with the expected performance penalty. If we use such animations in our applications, we need to be aware of this difference.

One distinction that straddles this divide, that is particularly important for mobile development, is the difference between physical handsets. Android phones have a physical back button, Apple devices just have a single Home button, Palm webOS devices have a swappable panel area, and so forth. Different devices have different screen sizes; some devices, such as Apple's iPhone 4, have unusual DPIs that affect the screen resolution.

We're going to use a variety of strategies to deal with these differences, by writing a code that is compatible across platforms.

Using a single codebase

We've seen in Chapter 2 how easy it is to copy a single application directory into multiple different projects. We're going to take a slightly different tack here, and use a symbolic link to have a single www directory on the file system connected to two platform-specific projects: an iPhone app, and an Android app.

We'll be targeting iOS 4.2 and Android 2.2 in this chapter, so make sure you have the latest versions of the SDKs at hand.

Since we'll have only one copy of the existing code on our file system, any changes we make will be propagated to all platforms, though we'll still have to build, and maybe clean, before seeing the changes.

There has been some work on the PhoneGap project on standardizing folder structures for cross-platform development; core-team developer Brian Leroux has a project called **Cordova**, available at `http://github.com/brianleroux/Cordova`, which attempts to solve some of these concerns. However, at the time of writing, these approaches are still very much in flux—your best bet is to manage your own file system in a manner you're comfortable with.

This technique only works for simple PhoneGap applications that do not use any of the native PhoneGap APIs, since each platform by default will build its own `phonegap.js` file into the shared www directory. We will look at more robust approaches later on.

Time for action - Detection and fallbacks

We're going to start a new application called **Red Green Refactor**! Red, Green, Refactor is a phrase drawn from the **Test-Driven Development**, a development approach we will studiously avoid throughout this book. But it does come in handy when you can use it.

1. Start by creating the www directory as usual, and three files: `index.html`, `app.js`, and `style.css`. Fill in `index.html` as follows:

```
<!doctype html>
<html>
    <head>
        <title>Red Green Refactor</title>
        <script src=»app.js»></script>
        <link rel=»stylesheet» href=»style.css»/>
    </head>
    <body>
        <div class=»screen» id=»red»>RED</div>
        <div class=»screen» id=»green»>GREEN</div>
        <div class=»screen» id=»refactor»>REFACTOR</div>
    </body>
</html>
```

2. And add the following to `style.css`:

```
.screen {
        width: 320px;
        height: 480px;
        float: left;
        color: white;
        position: absolute;
        font-size: 50px;
        text-align: center;
        top:0px;
    }

    .screen#red         { left: 0px; background: red; }
    .screen#green       { left: 320px; background: green; }
    .screen#refactor    { left: 640px; background: blue; }
```

3. Don't worry about `app.js` for now. Let's open what we have in Safari:

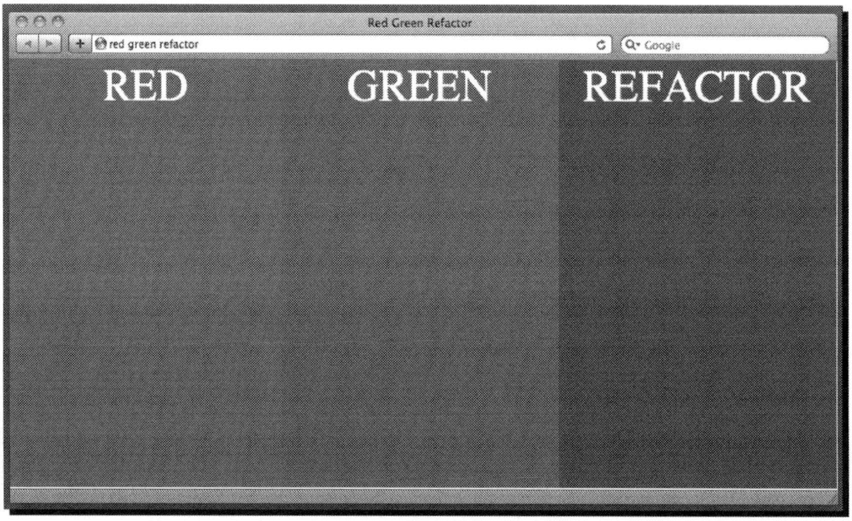

4. Cool, we can see our three states. Now let's fix their behavior: we want to display one state at a time, and switch between them on the user's action. Add the following to `app.js`:

```
var screens = ['red', 'green', 'refactor'];

    document.addEventListener('click', function (evt) {
        if (evt.target.getAttribute('class') == 'screen') {
            var oldScreen = evt.target.id;
            var newScreen;

            screens.forEach(function (screenId, i) {
                if (screenId == oldScreen) {
                    if ((i+1)<screens.length) {
                        newScreen = screens[i+1];
                    } else {
                        newScreen = screens[0];
                    }
                }

                document.getElementById(screenId).style.display =
                'none';
            });

            document.getElementById(newScreen).style.display =
            'block';
        }
    });
```

5. Each time the user taps on the screen, the display switches to the next one in the cycle. To keep things visible on a mobile device, let's make a couple of changes to the stylesheet:

```
.screen {
        width: 320px;
        height: 480px;
        float: left;
        color: white;
        position: absolute;
        font-size: 50px;
        text-align: center;
        top: 0px;
        left: 0px;
        display: none;
 }

    .screen#red          { display:
     block; background: red; }
    .screen#green        {
     background: green; }
    .screen#refactor     {
     background: blue; }
```

6. You can see that we've removed the left offset from each individual screen, set the screens to be hidden by default, and set the red screen to be initially visible. Reload `index.html` in Safari and check out the changes:

7. So that's the basic logic. Now, we want our application to look a bit snazzier, so let's use the aforementioned **CSS transforms** to display each new screen. The logic here is a bit more complicated, but you should be able to follow along: what we want is for each new screen to force the earlier one to slide off the screen.

8. First, we're going to modify the CSS. Note that instead of setting the element's display property, we're just setting CSS3 transition and transform properties (`-webkit` prefixed, as is standard for beta CSS3 implementations):

```
.screen {
    width: 320px;
    height: 480px;
    float: left;
    color: white;
    position: absolute;
    font-size: 50px;
    text-align: center;
    top: 0px;
    left: 0px;

    -webkit-transition-duration:
     600ms;
    -webkit-transition-property:
     translate;
    -webkit-transform: translate
     (-320px,0px);
}

.screen#red          {
    -webkit-transform:
     translate(0px,0px);
    background: red;
}
.screen#green        { background: green; }
.screen#refactor     { background: blue; }
```

9. And let's modify our click handler in `app.js` also:

```
document.addEventListener('click', function (evt) {
        if (evt.target.getAttribute('class') == 'screen') {
            var oldScreenEle = evt.target;
            var oldScreen = oldScreenEle.id;
            var newScreen, newScreenEle;

            screens.forEach(function (screenId, i) {
                if (screenId ==
                oldScreen) {
```

```
                    if
        ((i+1)<screens.length) {

                        newScreen = screens[i+1];
                    } else {
                        newScreen = screens[0];
                    }

                    newScreenEle =
                    document.getElementById(newScreen);
                }
            });

            newScreenEle.style.webkitTransform =
            'translate(0px,0px)';
            oldScreenEle.style.webkitTransform = 'translate(-
            320px,0px)';
        }
    });
```

10. Now to go back to Safari and start clicking—you can see a smooth back-and-forth motion on each transition. With a little more work we could implement a sliding motion, but the bidirectional effect is one that I'm fond of, and suits the repetitive nature of the red-green-refactor cycle.

11. The next stage is to set up Red Green Refactor as a PhoneGap-iPhone project (we'll be setting up an Android project soon also, so you should start loading the emulator now). Set up the iPhone project through Xcode as in the previous chapters, and modify the project on the file-system to point to our `www` directory (please see Chapters 1 and 2 if this part is unclear). Launch the app in the simulator now, and verify that the transitions are silky smooth.

12. Notice that there are no transitions. It happens that there's a bug in event delegation on the iPhone—events don't bubble up as far as the body element, or the document, unless an explicit click handler is defined on each element that receives the clicks. The bug itself has been described by JavaScript guru Peter-Paul Koch at `http://www.quirksmode.org/blog/archives/2010/09/click_event_del.html`; we're going to add the following JavaScript to work around it:

```
document.addEventListener('DOMContentLoaded', function () {
    function emptyClicker() {};

    screens.forEach(function (screenId) {
        document.getElementById(screenId).
addEventListener('click',
emptyClicker);
    });
});
```

13. Once the application is loaded, we loop through the list of screens and assign an event handler (the `emptyClicker` function) to each element's click event.

14. This is quite trivial since we only have three divs, and none are dynamically created or removed at runtime. The bug is a lot nastier when we're doing more dynamic work, with a larger number of elements to work on—we need to keep track of every new element that gets created, and manually add an event handler. At any rate, here's how the patched application looks in the iPhone simulator:

And the transitions look pretty good themselves.

15. Okay, now on to Android. Again, refer back to earlier chapters to create a new PhoneGap-Android project using droidgap, and copy our `www` directory into that app project. Now spin up the emulator and install:

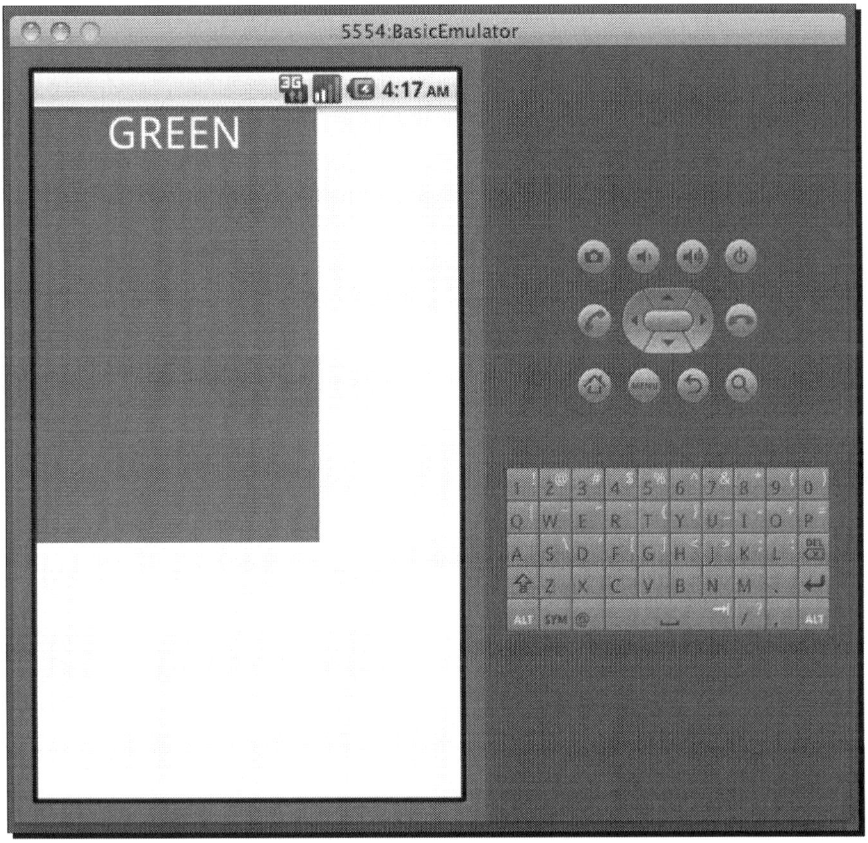

16. Well... it's not great, is it? Not only are the transitions slow and jerky, but the viewport is the wrong size (with this emulator—different Android Virtual Devices may see different results). Not too clever at all.

17. Let's fix the viewport first, since it's an easy change to make. Add the following tag to the `<head>` of our `index.html`:

```
<meta name="viewport" content="initial-scale=1.0,width=device-
width,user-scalable=no" />
```

That probably should've been there from the start, but never mind. Reload it in the Android emulator, and things look a bit better:

18. That's much better, but not perfect (again, on the Android Virtual Device that I am using); since Android screen dimensions and aspect ratios vary from device, we have to be careful when specifying absolute width and heights. Unfortunately, CSS transitions require those values to be set, as well as absolute positions, for animations to work correctly. We can work around this by setting those values absolutely at runtime, when we can detect their correct values, but let's not get too far ahead of ourselves.

19. Our next goal is to fix those wonky transitions on Android. As mentioned earlier, the issue is that certain platforms hardware accelerate 3D transforms, while others do not, while supporting the same CSS features and JavaScript APIs.

20. Because of this, we can't feature detect the difference—we'll need to do some browser sniffing. We need to check what the `userAgent` of the browser is, attempt (well, guess) whether or not hardware acceleration is supported, and store that as a flag we can fork behavior on. Here's how `app.js` ends up looking:

```
var screens = ['red', 'green', 'refactor'];
var hasHardwareAcceleration = false;

document.addEventListener('click', function (evt) {
    if (evt.target.getAttribute('class') == 'screen') {
        var oldScreenEle =
        evt.target;
        var oldScreen =
        oldScreenEle.id;
        var newScreen,
        newScreenEle;

        screens.forEach(function (screenId, i) {
            if (screenId ==
            oldScreen) {
                if ((i+1)<screens.length) {
                    newScreen = screens[i+1];
                } else {
                    newScreen = screens[0];
                }

                newScreenEle = document.getElementById(newScreen);
            }
        });

        if (hasHardwareAcceleration) {
            newScreenEle.style.webkitTransform =
            'translate(0px,0px)';
            oldScreenEle.style.webkitTransform = 'translate(-
            320px,0px)';
        } else {
            newScreenEle.style.display = 'block';
            oldScreenEle.style.display = 'none';
        }
    }
});

document.addEventListener('DOMContentLoaded', function () {
    hasHardwareAcceleration =
    !!(navigator.userAgent.match(«iPhone»));
 function emptyClicker() {};
```

```
        screens.forEach(function (screenId) {
            var screenEle = document.getElementById(screenId);
            if (!hasHardwareAcceleration) {
                screenEle.style.webkitTransform =
                  'translate(0px,0px)';
                if (screenId != 'red') {
                    screenEle.style.display = 'none';
                }
            }
            screenEle.addEventListener('click', emptyClicker);
        });
    });
```

21. Please note the expression `!!navigator.userAgent.match("iPhone")`—we're using two Boolean NOT operators (`!`) to coerce the result of the `match` function into a Boolean value (`true` or `false`).

22. You can see what we're doing here—we check if the user agent contains iPhone, and, if it doesn't, we fall back to our older behavior (showing and hiding divs).

23. Phew. Now rebuild for both iPhone and Android and verify that the transitions look acceptable on both platforms. Looks good to me.

What just happened?

We took our first stab at branching our code execution based on the targeted platform, using the least sophisticated method possible—querying `navigator.userAgent` for an expected result. We'll examine why that's a bad idea in a second but first, let's go over what happened in that last step.

User agent sniffing

The last block of code in `app.js` is an event listener bound to the `DOMContentLoaded` event—in effect, the first thing that's executed when our application is loaded:

```
document.addEventListener('DOMContentLoaded', function () {
    hasHardwareAcceleration = !!(navigator.userAgent.match("iPhone"));
    function emptyClicker() {};

    screens.forEach(function (screenId) {
        var screenEle = document.getElementById(screenId);
        if (!hasHardwareAcceleration) {
            screenEle.style.webkitTransform = 'translate(0px,0px)';
            if (screenId != 'red') {
                screenEle.style.display = 'none';
```

```
                }
            }
            screenEle.addEventListener('click', emptyClicker);
        });
    });
```

The first thing we do is query `navigator.userAgent`. Like JavaScript's `alert` and `eval` functions, `navigator.userAgent` is an especially lo-fi tool that should be used with extreme prejudice in most circumstances. However, mobile development has a way of eroding those prejudices, and leaving you to pick up the least desirable tools. Here's that value on my iPhone Simulator:

> *Mozilla/5.0 (iPhone Simulator; U; CPU iPhone OS 4_1 like Mac OS X; en-us) AppleWebKit/532.9 (KHTML, like Gecko) Mobile/8B117*

The reasons why this is a horrible thing to work with:

- **There's a lot of information we don't need**: The WebKit build number, for instance, or whatever 8B117 stands for.
- **There's a lot of inaccurate information**: The phrases "Mozilla", "like Mac OS X", and "like Gecko" are all misleading—it is not a Mozilla browser, it is not running on Mac OS X, and the rendering engine is WebKit, not Gecko. If we're just looking for the presence of those strings, we will be misled.
- **It's not very well-structured**: We can't say `navigator.userAgent.OS`, for instance. It's just a long string.

PhoneGap exposes an object called device that solves the above issues—you can access `device.platform` to get the OS name, for instance, and `device.version` to get the version—but that doesn't solve the underlying problem: when you browser sniff, you're relying on a single variable to tell you about another, independent variable.

If you have a network connection, one option for dealing with user agent strings is to use **BrowserScope**, "a community-driven project for profiling web browsers". BrowserScope will take a user agent string you pass it and tell you the browser, operating system, version, and much more. Although it's not suitable for our current purposes, where we need the application to work identically offline, BrowserScope is a very useful tool. You can find more information at `http://www.browserscope.org/`.

In our example, we check for a match against **iPhone** since that seems to work for mobile devices—but note that we lose the transitions when viewing our page in desktop Safari, since it doesn't match iPhone.

At any rate, if we infer that hardware acceleration is not present, we remove the transitions styles and hide the inactive screens.

The other change is when a click event occurs—it's fairly clear what's happening:

```
if (hasHardwareAcceleration) {
   newScreenEle.style.webkitTransform = 'translate(0px,0px)';
   oldScreenEle.style.webkitTransform = 'translate(-320px,0px)';
} else {
   newScreenEle.style.display = 'block';
   oldScreenEle.style.display = 'none';
}
```

That is the basic mechanism when we branch platform code based on a runtime variable—detect the (typically unchanging) value of that variable once, at runtime, and use the result to branch the code whenever appropriate.

Feature detection

As we've stressed, sniffing the user agent is a sub-optimal option for branching code execution. The preferred approach, for PhoneGap and web development in general, is **feature detection**: testing for the existence of a particular object, or a property on an existing object, and forking the code on that basis.

Here's a simple example of the first kind, checking if the global `WebSocket` object is present in the DOM—if it's not, we can fall back to using a standard `XMLHTTPRequest`:

```
if (!!window.WebSocket) {
    var socket = new WebSocket('ws://sample.com:111/updates');
    // do some fun socketing
} else {
    var fakeSocket = new XmlHttpRequest();
    // do some fun fake socketing
}
```

If web sockets are not available in the current environment, `window.WebSocket` will evaluate to `undefined` and `!!window.WebSocket` will be `false`.

This approach to feature detection is very useful for seeing if new HTML5 JavaScript APIs are present in the current environment—we can test for the presence of local storage, web workers, and even PhoneGap APIs using this technique. For CSS3 properties, we typically need to test for the presence of an attribute on a DOM element object. Here, for example, is a test for the CSS3 transform property:

```
var ele = document.createElement('div');
if (typeof ele.style.webkitTransform == "string" || typeof ele.style.
transform == "string") {
    // transforms are available
} else {
    // transforms are not available
}
```

We create a new empty `div` element and check its `style` object for either a `webkitTransform` property or a `transform` property. If either one is present as a string, we know that CSS3 transforms are available.

The eagle-eyed among you will have noticed that the above code will not detect vendor-prefixed CSS transforms in Opera, Firefox, or Internet Explorer user agents, but it would be the same basic principle (check for `oTransform`, `mozTransform`, or `msTransform` respectively).

If all of this sounds like a lot of work, there are thankfully a large number of open source JavaScript libraries that take care of feature detection. The most popular of these is **Modernizr**, developed by Faruk Ates and Paul Irish, which adds CSS classes to your document based on supported browser functionality. This allows you to take advantage of the feature detection goodness without having to write any JavaScript to take care of it. Modernizr is available at `http://www.modernizr.com`.

There is also a pure feature-detection library available called `has.js`, developed by Pete Higgins and a number of other JavaScript luminaries. `Has.js` is essentially a better-written and more robust version of what we've done in our example; reading the source code to the library gives a great overview of how feature detection works. It can be found at `http://github.com/phiggins42/has.js`.

Pop quiz – Feature detection versus UA sniffing

1. For each of these examples, would you be able to use feature detection or would you have to fall back to user agent sniffing?
    a. Checking if the device has a hardware back button.
    b. Finding support for a WebSQL SQLite database.
    c. Finding support for multitouch events.
    d. Checking the total memory available to your application.
2. When performing feature detection, when should we execute the detection code?
    a. As soon as the page renders (in an inline script tag).
    b. On the `DOMContentLoaded` event.
    c. At the time we need to access that feature.
3. What kind of code branching should you do with user agent sniffing?
    a. OS level changes: one thing for Android, another for iOS.
    b. Device-level: one thing for the Samsung Galaxy Tab, another for the HTC Dream.
    c. Build-level changes: one thing for WebKit builds below 532, and another for 532+.

Media queries

For certain classes of platform/device inconsistencies, **CSS Media Queries** can be very useful. Seasoned web developers will be familiar with the media property on CSS link tags, which allow different stylesheets to target different media. Here is a simple example of one CSS file being used for on-screen rendering, and another for printing:

```
<link rel="stylesheet" media="screen" href="screen.css" />
<link rel="stylesheet" media="print" href="print.css" />
```

With CSS3 capable browsers, we can do much more interesting things with media queries. One thing we noticed in our above example was that the screens on our Android emulator did not take up the full space available to the application, unlike those on the iPhone simulator.

If you remember, we set the height and width of our `screen` class absolutely in order to ensure our CSS transforms behaved correctly on all iOS devices. Since the Android version won't have transforms anyway, let's look at how we can use the dimensions of the device to set the width and height.

First, let's change the width and height to percentage values in our stylesheet, rather than pixel values:

```
.screen {
    width: 100%;
    height: 100%;
    ...
}
```

This is going to be our default value—if we resize our Safari window, we can see this in action:

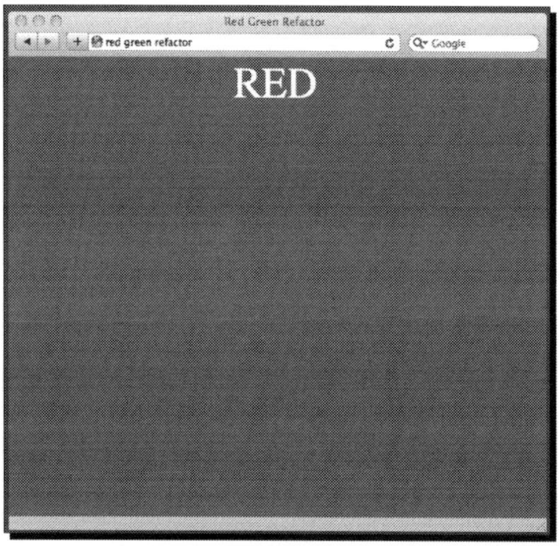

Now we can add a media query to the stylesheet to check if the screen matched the iPhone dimensions—if so, we'll resize the screens, with a nice surrounding margin:

```
@media all and (max-device-width:320px) and (max-device-height:480px)
{
    .screen {
        margin: 10px;
        width: 300px;
        height: 440px;
    }
}
```

Let's rebuild our iPhone application and check that the changes have been persisted:

Great! And if we reload in Safari, or rebuild our Android project, we can see that those have been unaffected by the new CSS.

You can test similar media queries—`max-device-width`, `min-device-width`, `max-device-height`, or `min-device-height`—on Android devices or virtual devices. Screen dimensions will vary, so be sure to set values appropriate for the device you're targeting.

A word of caution: like other new HTML5 and CSS3 features, media queries have limited support; especially the more advanced ones (for example, querying the device pixel-density or the device orientation). Using a library like Modernizr will help detect exactly what is available—it can be found at `http://modernizr.com`.

Have a go hero

Our application has only been styled for portrait orientation—in landscape mode, it looks less clever:

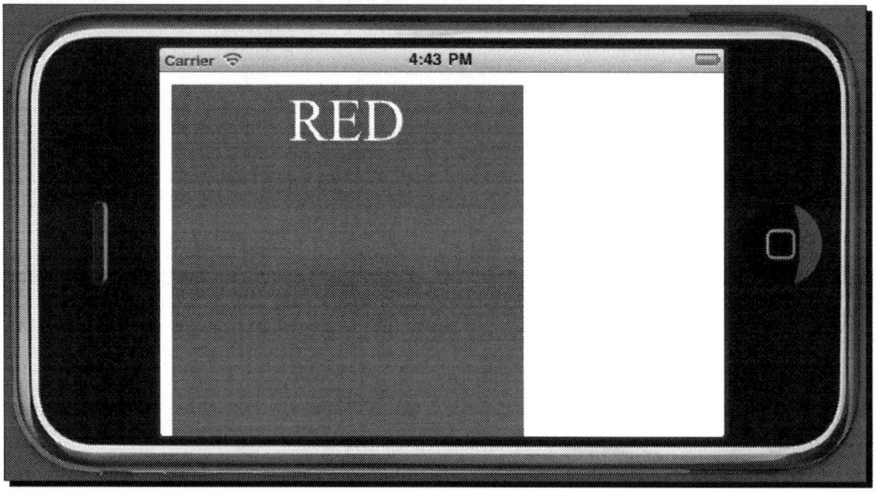

How can your media queries accommodate the landscape view appropriately? How will this affect the CSS transforms that are applied? Will any changes need to be made to fix things on the Android project? Can we make all the changes through CSS, or will we need to write some JavaScript as well?

Preprocessing code

Let's close the chapter with one last approach: preprocessing the JavaScript, HTML, and CSS that constitutes your application, so it is precisely built for each platform you target.

PhoneGap has the advantage over web development of running in an essentially safe environment—you know beforehand which platforms you will be targeting, you can develop and test on a (relatively) broad swathe of the devices you're interested in, and you can make several assumptions about the browser environment that developers for the Web at large cannot.

For example, instead of having a single `application.js` file with sections intended for different devices that looks like this:

```
if (iPhone) {
  doIphoneThing();
} else if (android) {
  doAndroidThing();
} else if (blackberry) {
  doBBThing();
}
```

There would be an `application.iphone.js` file containing just `doIphoneThing()`, and equivalent `application.android.js` and `application.blackberry.js` files. Some logic at build-time would select which file gets included for each platform.

With these welcome constraints, you do have the option of munging together your code before building to a device. There are as many approaches for this as there are PhoneGap developers, but here are some good rules of thumb:

- **Use an environment you're familiar with**: Some developers are most comfortable writing shell scripts, some like writing Java, some like writing Ruby, and so on and so forth. HTML, JavaScript, and CSS are all just plain-text formats, and any environment that can write to the file system is capable of being used for pre-processing.
- **Share what you can**: One common approach is to have a JavaScript file for every platform—`application.js`—and separate files for platform-specific code—`app.android.js`, `app.iphone.js`, and `app.blackberry.js`. This is similar to PhoneGap itself, where the same APIs are implemented in different fashions on different platforms, and is an easy method to get up and running.
- **Don't go overboard**: On one project, I tried to take this approach a bit too far, by storing every JavaScript, CSS, and HTML file as a Mustache template that is generated to platform specific code at build time. This helps when you need to share the vast majority of your code, but becomes unwieldy quickly. Try to separate out code that is separately targeted, as far as possible.

Summary

In this chapter, we took aim at the biggest concern for PhoneGap developers: differences in functionality between the platforms that we target. Here are the lessons we learned:

- In many cases, we can use feature detection to safely find the presence of features/APIs we wish to use. This is the optimal approach in situations where it is available, and libraries like Modernizr let us take advantage of the approach painlessly.
- For many UI purposes, CSS3 media queries allow us to flexibly and effectively tweak layouts based on the runtime environment.
- Using the device's platform/OS, either through PhoneGap's `device` object or through the browser's `navigator.userAgent` property, is a brittle and unreliable option that we have to fall back on, too much for comfort.

There are no perfect approaches for these purposes—we're always dealing with fragile tools that have to be backed up by manual testing and attention. Developing for the fast-moving mobile platforms requires you to be aware of the trade-offs and considerations that you're making with your tools. That said, we can go an awfully long way by using these techniques, with a bit of caution.

In the following chapters, we'll dive deeper into the APIs and interfaces available, either at the browser-level or exposed by PhoneGap, to create outstanding applications. We'll need the techniques from this chapter to ensure graceful degradation when our applications execute in less privileged environments.

5

HTML5 APIs and Mobile JavaScript

As PhoneGap fundamentally relies on the capabilities of each supported device's native web browsing implementation, it's important to be aware of what these capabilities are. Browser vendors have pushed each other to increasingly support HTML5 APIs—new JavaScript interfaces that expose powerful functionality to developers. Taking these APIs, along with some best practices around using JavaScript on mobile devices, allows us to get the most out of our applications.

In this chapter, we will see how to take full advantage of JavaScript on mobile devices, and the new HTML5 APIs, by:

- Utilizing the mobile-optimized XUI library to write smaller, more efficient JavaScript code
- Exploring the HTML5 media elements and their JavaScript APIs
- Dynamically creating graphics with the HTML5 canvas element
- Seeing whatever HTML5 capabilities are in the pipeline for mobile devices

Mobile JavaScript

A fair question arises at this point: what is mobile JavaScript, as distinct from JavaScript qua JavaScript?

It's a tricky question to answer. In terms of syntax and semantics, except outliers like BlackBerry devices below OS 5.0, you can expect all of the language features of the ECMAScript language third edition (the most widespread version of JavaScript at the time of writing, on any platform). As with desktop browsers, support for ECMAScript 5 (the latest standardized version of the language) is inconsistent—we're going to stick to ECMAScript 3 in this book, for this reason.

Where mobile JavaScript does distinguish itself in the constraints placed on the language runtime, due to the characteristics of the hardware itself. Even in 2011, all mobile devices are considerably underpowered and limited compared to their desktop counterparts. This affects the memory that is available to your application and the speed your code will execute.

Most of the time this constrains what you can do in your application, though your mobile browser does most of what your desktop browser can do, it does it slower and less effectively. Because of this, there's a simple rule to follow with mobile JavaScript: write less code!

XUI

XUI is a little JavaScript library originally written by Rob Ellis, one of the founding fathers of PhoneGap, and maintained by Brian Leroux, one of the core members of the PhoneGap team. XUI mirrors the API of the popular **jQuery** JavaScript library—CSS selectors playing a central role, chained function calls, terse syntax—while jettisoning much of the feature set and browser support of jQuery itself.

In a public-facing website, you want your code to run effectively no matter which browser the user has at hand. Therefore, libraries such as jQuery (to their great credit) are filled with the code designed to degrade gracefully on all major browsers, while providing a consistent API.

In a PhoneGap application, you know at build-time which platforms your application will run on, and what the capabilities of the web view rendering your application are. This allows XUI to take advantage of newer browser features, which reduce the overall quantity of code.

One classic example is jQuery's selector engine, **Sizzle**. Sizzle is a widely compatible library that takes advanced CSS selectors and returns collection of DOM elements. Over the past few years, web browsers have introduced a native function called `document.querySelectorAll` that does essentially the same thing. XUI assumes this function will be available (or it can be built with Sizzle included); jQuery doesn't make this assumption, and so all clients receive Sizzle as part of the jQuery library.

Time for action – Downloading, building, and using XUI

We're going to go back to the food listing application we last touched on back in Chapter 3, and see how we can use XUI to simplify the code and quickly add some extra functionality. Firstly, we need to get our own copy of XUI:

1. Get your own copy of XUI—the download is available at `http://xuijs.com/downloads`.
2. Copy the new JavaScript file into your food project directory, and add a script tag to include XUI, just before the first, large script tag:

```
<script src="xui.js"></script>
```

3. Let's take some of the existing JavaScript code and use XUI's API rather than the native DOM commands to do the same thing. One thing that XUI does, like a lot of JavaScript libraries, is allow terse selection of DOM nodes, and event bindings. We can change the following line:

```
document.getElementById("foodForm").addEventListener("submit",
function (evt) {
```

To the following:

```
x$("#foodForm").on("submit", function (evt) {
```

This will get the same effect, with almost half of the characters removed. Instead of calling `document.getElementById`, as we have done throughout the code in the sample application, using the `x$` function allows us to utilize more complex CSS selectors, and have more feature-rich objects returned.

4. Let's try to use one of these complex CSS selectors. When we add a new food, we're going to display its name as bold, and in white. Then, after a second, we'll restore it to the normal color.

5. Find the `addNewFoodItem` function, which looks like this:

```
function addNewFoodItem(foodName) {
  var newFoodItem = document.createElement('li');
  newFoodItem.innerHTML = foodName;
  foodList.appendChild(newFoodItem);););

  aFoodDetail.foodName = foodName;
  renderOurTemplate(aFoodDetail, function (markup) {
    document.getElementById(«foodDescription»).innerHTML =
markup;;
  });
}
```

6. We're going to focus on the first three lines of the function for now. Firstly, we can make the process of adding the new food item to the DOM a lot terser using XUI—we know how the `li` element should look, so we can insert that directly, changing those three lines to:

```
x$('#foodList').bottom('<li>' + foodName + '</li>');
```

7. Note the use of the `bottom` function: XUI has a number of helpful functions for adding new markup relative to the selected element (you can find all of them documented here: `http://xuijs.com/docs/dom`). Another useful one is `inner`—we can use that in the `renderOurTemplate` callback:

```
renderOurTemplate(aFoodDetail, function (markup) {
  x$(«#foodDescription»).inner(markup);
});
```

8. The next step is to select the last element of the list, so that we can style it. We're going to use the CSS `last-child` pseudo-selector to get the last item in the list of foods. Open Web Inspector in Safari to test this out; enter the following:

```
x$('li:last-child')[0].innerHTML
```

9. The selector finds all of the DOM nodes that match our query, and we then use the array index syntax to grab the first DOM node specifically (in this case, we want to check the innerHTML attribute to make sure the correct item has been selected).

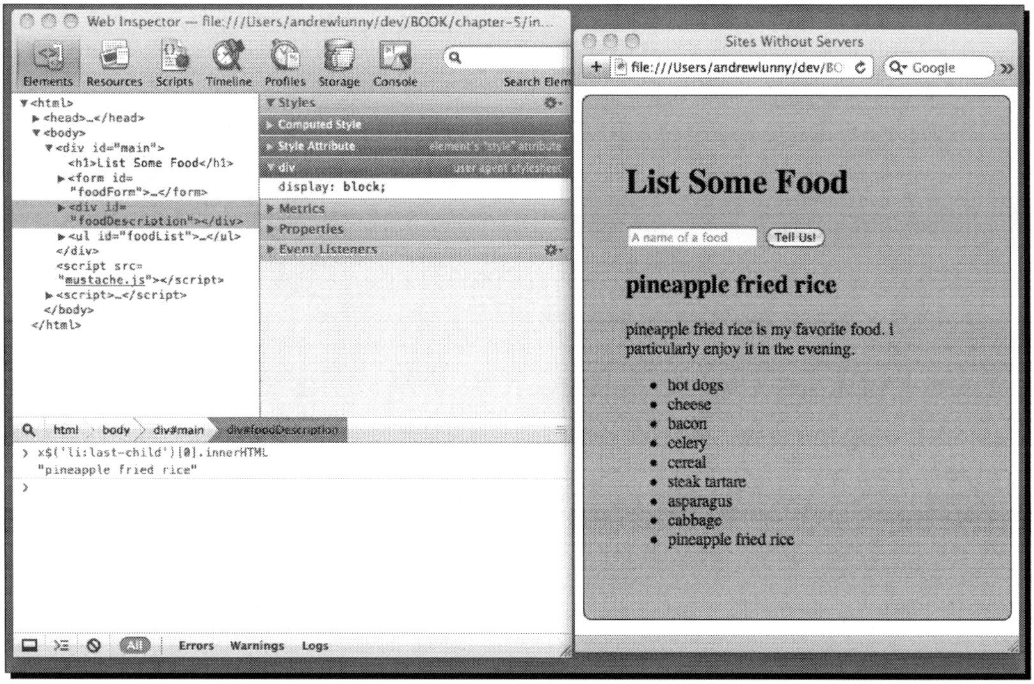

In this case the last element is **pineapple fried rice**—it'll differ depending on what you've entered into your list.

10. The next step is to style that element after it's displayed. XUI gives us a syntax that, again, is terse and easy to read. Here is our modified `addNewFoodItem` function that highlights the last item in the list:

```
function addNewFoodItem(foodName) {
    x$('#foodList').bottom('<li>' + foodName + '</li>');
    x$('li:last-child').css({
        'font-weight': 'bold',
        'color': 'white'
    });
```

```
  aFoodDetail.foodName = foodName;
  renderOurTemplate(aFoodDetail, function (markup) {
    x$(«#foodDescription»).inner(markup);
  });
}
```

Let's give that a shot in Safari and see how it looks:

11. Okay, `addNewFoodItem` is running in a loop, and each time an item is added, it is the last child. Also, every time we highlight the last child, we're not touching the former last child. This is clearly not what we want.

12. We need to clear the styles that were previously set. The simplest way to this in XUI is to call the `css` function again, this time with empty values for the fields that were affected. We're going to create a JavaScript object called `defaultStyle` that holds these empty values:

```
var defaultStyle = {
    'font-weight': '',
    'color': ''
}
```

13. Then, we'll apply `defaultStyle` to all of the `li` elements after the page has initially been populated—add this code to the top of the `addNewFoodItem` function:

```
var storeLength = window.localStorage.length;

for (var i; i < storeLength; i++) {
  storedFoodName = window.localStorage.key(i);
  if (storedFoodName.match(/^food[.]/)) {
      addNewFoodItem(window.localStorage.getItem(storedFoodName))
    }
}

x$('li').css(defaultStyle);
```

14. Let's reload in Safari, and add a couple of foods to be sure it still works:

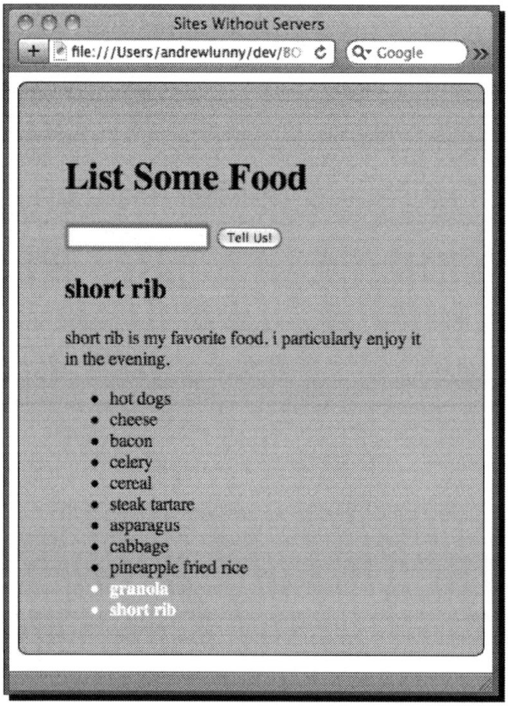

15. We're making progress, but adding multiple foods results in multiple foods being highlighted, which is not what we want. That's an easy fix though—simply set the older children to the default style every time we highlight a new one:

```
x$('li:last-child').css({
        'font-weight': 'bold',
        'color': 'white'
    });
x$('li:nth-last-of-type(n+2)').css(defaultStyle);
```

16. Once more, we're using a fancy CSS3 selector; this time the `nth-last-of-type` property allows us to select all but the last `li` element.

    Now each newly entered food is highlighted:

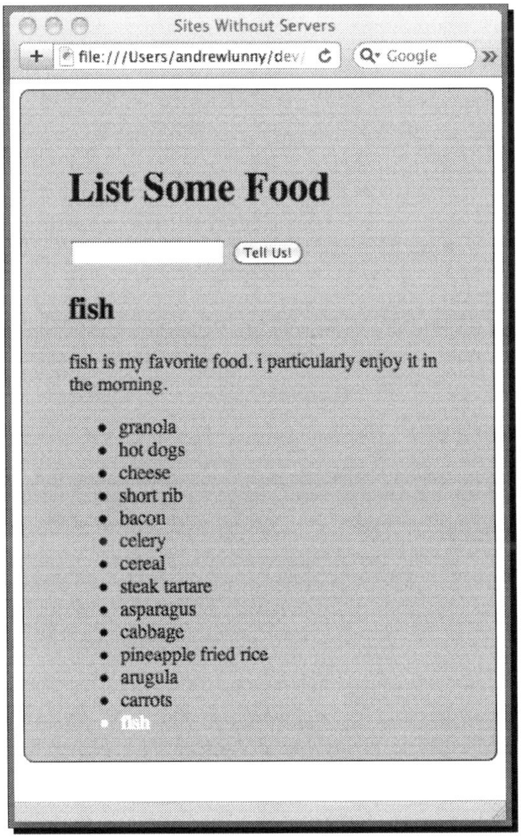

17. Things are going well, but our list is starting to get unwieldy at this point—we would like to remove some of the items. Firstly, we're going to have to add a delete button for every item on our list—we're going to use a CSS trick to do this. Firstly, in the style element at the top of the page, let's define our `deleter` class:

```
.deleter:after {
    color: red;
    font-family: sans-serif;
    font-size: 10px;
    content: 'X';
}
```

18. Eagle-eyed readers will have noted that this isn't a `deleter` class at all, but a `deleter:after` pseudo-class. Using this class gives us access to the content property, so that text content (with no textual value) can be added just through CSS. Now we just need to add that dummy element to each list item that we add, changing the first line of `addFoodItem` to the following:

```
x$('#foodList').bottom('<li>' + foodName +
                          ' <span class=»deleter»></span></li>');
```

19. This displays little red marks next to each item. They don't do anything yet, but we're working on that:

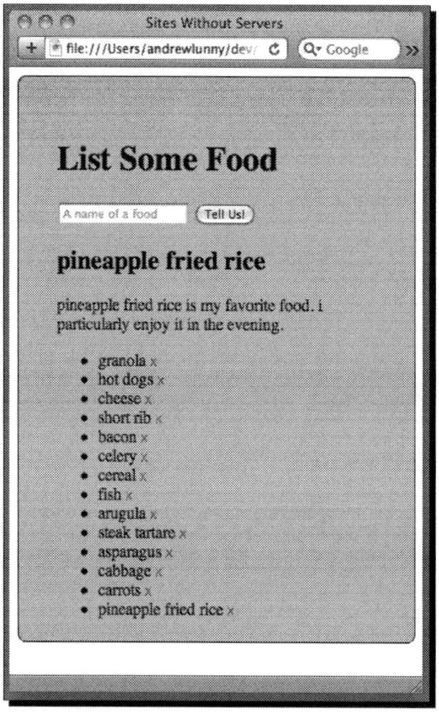

20. Removing the elements from the DOM is a doddle with XUI. You'll remember there's some event delegation code at the bottom of our `index.html` page, for remotely calling to the Twitter API. Since these delete buttons are dynamically added too, we'll need to use event delegation to ensure that they are always working.

21. The previous event delegation code was essentially one big `if` statement—if the target element is a list item, make a request to the Twitter server. We just need to add an `else` clause to that statement; if the target element is a `deleter`, remove its parent (the list item) from the DOM:

```
if (evt.target.tagName == "LI") {
  // all of the LI-specific code
} else if (x$(evt.target).hasClass('deleter')) {
    x$(evt.target.parentNode).remove();
}
```

I've highlighted the important line: grab the parent element and call `remove`.

22. Just one more thing—we'll need to remove the item from `localStorage` as well, otherwise it will show up again when we reload the page. This requires a few quick changes. Firstly, everywhere `addNewFoodItem` is called, we need to pass the key that the item is stored under in `localStorage`, for easy access. This is easy enough, since we have a uniform setup for the keys themselves.

There are two places this function is called: in the initial loop to populate the page, and in the submit handler for the form to add a new food. We can change the loop to pass the food key along with the item itself:

```
if (storedFoodName.match(/^food[.]/))
    addNewFoodItem(window.localStorage.getItem(storedFoodName),
    storedFoodName);
In the submit handler, we already have new key stored in
a variable, so it's even easier - just modify the call to
addFoodItem:
var newFood = foodField.value;
var foodKey = «food.» + (window.localStorage.length + 1);

addNewFoodItem(newFood, foodKey);
```

23. Now let's edit `addNewFoodItem` to take advantage of this new variable; all we need to change are the function's definition and its first couple of lines:

```
function addNewFoodItem(foodName, foodKey) {
    x$('#foodList').bottom('<li id=»' + foodKey+ '»>'
        + foodName + ' <span class=»deleter»></span></li>');
```

Since we're not doing anything else with the `id` field, we can hijack it to save the storage key.

24. Now just delete the item from `localStorage`, using this key. It's a quick change to the `deleter` event handler that we just defined:

```
} else if (x$(evt.target).hasClass('deleter')) {
    var listIem = x$(evt.target.parentNode);
    var storageKey = listIem[0].id;

    listIem.remove();
    window.localStorage.removeItem(storageKey);
}
```

25. Let's give it a shot; delete some foods, reload the page, and we should have success!

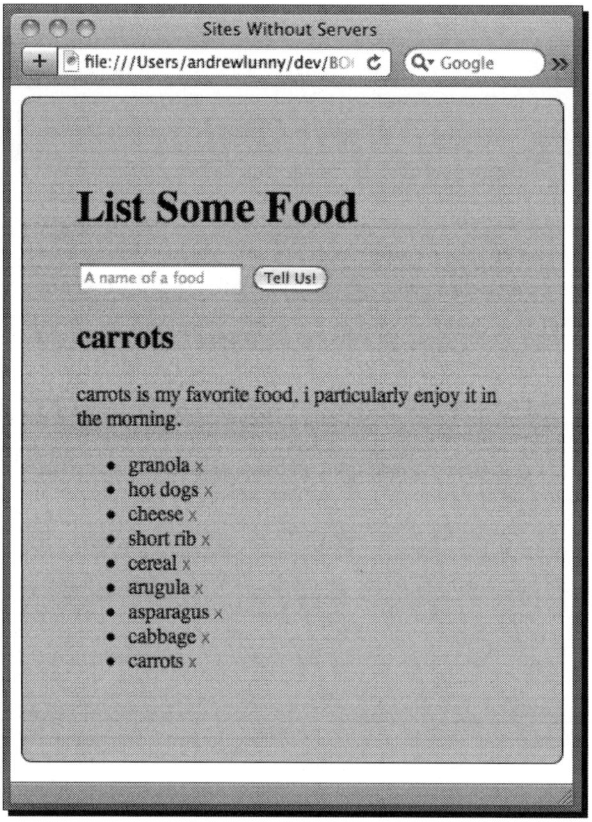

What just happened?

We had a whirlwind tour of XUI, investigating its DOM manipulation, event handling, and CSS editing functionality.

XUI has a restricted feature set, but the features that are available are the ubiquitous ones needed for every web-based application: those listed above, and also robust Ajax support. In the wider world of web development, libraries such as XUI have been omnipresent for some time now, whether Dojo, Prototype, jQuery, or any other have been in fashion. The benefits are clear: a terse syntax allows for less application-specific code, making the application easier to manage and extend.

Why not jQuery?

If we take it that a JavaScript library is important for your application development, why would you not want to use the most popular one on the web? Most web developers already have experience using jQuery in their day to day lives—why would they not use it for PhoneGap development?

Well the first answer is yes, you could use jQuery—it runs well enough on mobile devices—and has the advantage that there are mobile user interface libraries that depend on it—**jQTouch** and **jQuery Mobile**, to name the two most popular.

What's important to note, though, is that jQuery, like most other JavaScript libraries, is designed for use on the public Web, which leads to certain design decisions. XUI is essentially jQuery designed for PhoneGap, which leads to different design decisions. In particular:

- Uncompressed file size, on the public Internet, is less important than the size of a compressed (**gzipped**) resource. Modern HTTP servers compress files before they are transmitted, which are decompressed by the client's browser. Libraries like jQuery are compressed in this way, by a factor of about three. With PhoneGap applications, the important scripts are already on the local device, so this kind of compression is irrelevant.
- Mobile devices have limited resources, in terms of available memory and raw performance. Because of this, uncompressed file size is very important; in a typical desktop environment, the uncompressed file size is as good as irrelevant (within reason), once it's passed over the network. XUI optimizes for file size: it's about a tenth of the size of jQuery, at the time of writing.
- Sites facing the public Internet do not know which clients they will encounter; any good general purpose library developer will do their utmost to ensure that things either work as expected, or fail gracefully, whichever browser requests their scripts. With PhoneGap, we know ahead of time what kind of environment our scripts will run in; this allows us to make a number of assumptions about the target environment.

Once these design considerations are taken into account, it's easy to see the differences between the two libraries: jQuery favors compatibility and comprehensiveness (there are a *lot* more functions available in jQuery), while XUI emphasizes file size and browser targeting.

> A final note: XUI is a small, not especially stable, project, with a handful of active developers (including the present author). jQuery is the most popular JavaScript library in the world. Again, this goes both ways—if you have a problem with XUI, it's easy to look into the code, make a quick change, and patch things up, whereas with jQuery, you can hop on the IRC channel and speak to a vast community of talented developers who will be happy to help.

And of course, all of the above applies to the other excellent DOM frameworks available to the modern web developer—it's not just about jQuery!

Pop Quiz – XUI

1. Instead of removing food items from our page, let's say we want to hide them, with the option of bringing them back later. What would be the best way of doing this (assume that we have selected the correct element, and `hidden` is a CSS class with `display:none` as a property)?
    a. `x$(evt.target.parentNode).hide()`
    b. `x$(evt.target.parentNode).addClass('hidden')`
    c. `x$(evt.target.parentNode).css({'display': 'none'})`
2. Is XUI compatible with jQuery plugins?
    a. Yes: XUI has identical syntax to jQuery. We just need to redefine the `$` variable to equal `x$`.
    b. No: XUI has a restricted feature set, and does not have identical behavior to jQuery.
3. What's the primary advantage XUI has over other JavaScript (DOM) libraries?
    a. Smaller file size
    b. Faster performance
    c. jQuery-compatible syntax

HTML5

We've already encountered HTML5 APIs in previous chapters and above, in the use of `localStorage`, so it may be worthwhile to recap exactly what HTML5 is. As its name suggests, HTML5 is the latest specification for HTML, and the first such specification with an emphasis on web applications, as opposed to marking up documents. The HTML5 specification process is ongoing—lots of new APIs are being proposed, revised, and dropped all of the time.

HTML5 has received a lot of hype, particularly from Apple, as an alternative approach to Adobe's **Flash** technology. Flash has been the de facto standard for rich client-side applications on the desktop web for over a decade but, for reasons both technical and political, Flash has attained less traction on the mobile web. Thankfully, some of the new APIs included in HTML5—particularly support for audio and video elements, and dynamic bitmap and vector graphics—offer developers an alternative to Flash that is well-supported and standardized, and quickly gaining tooling and mindshare.

Unlike Flash, HTML5 is a set of generally discrete ideas and implementations—the HTML5 specification is not even finalized yet. Some HTML5 technologies, such as access to geolocation, are well-defined and well-supported across all of the first class mobile platforms. Some are in a state of flux—the Web Sockets proposal, at the time of writing, is currently being revised to fix security concerns with the underlying protocol. Using a tool such as the previously mentioned Modernizr should give a good idea of which features are accessible on each device you target—feature detection and manual testing are your friends here.

HTML5 is especially important in PhoneGap development since the richness of its available interfaces allows applications built on the mobile web to match, and even exceed, the quality of native applications. Let's see how we can start to incorporate some of those.

Media elements

The most prominent "killer apps" of HTML5 have been the media elements: the `audio` tag and the `video` tag. While older browsers required a third-party plugin (most commonly Flash) to play inline audio and video in web applications, HTML5 compliant browsers can embed audio and video files directly, in the same way the `img` tag works.

This tutorial won't cover encoding video or audio for mobile devices, or the morass of codecs and formats to be aware of. One excellent resource on this topic is Mark Pilgrim's Dive Into HTML5 site—his chapter on web video, `http://diveintohtml5.org/video.html`, covers this ground in superb detail. We'll be using an **H.264** encoded MP4 file—H.264 is the best supported video codec on current mobile browsers.

Time for action – My dinner with PhoneGap

To get a sense of how we can use the `video` tag and its associated JavaScript API, we're going to extend our food application slightly—if the user enters a fruit in the form, the normal behavior will be skipped, and we'll play the video:

1. For this example, we'll need to know whether the entry is a fruit or not. Let's write a simple function to check the entry against our list of fruit:

```
function isFruit(foodName) {
    var fruits = ['apple','orange','peach','raspberry'];
    var i = 0;

    for (i; i<fruits.length; i++)
        if (fruits[i]==foodName) return true;

    return false;
}
```

2. Now let's tweak the form's submit handler so that it doesn't write the fruit to the list or save it to local storage. For now, we'll alert the name of the fruit with a message:

```
x$("#foodForm").on("submit", function (evt) {
  evt.preventDefault();
  var newFood = foodField.value;
  var foodKey = «food.» + (window.localStorage.length + 1);

   if (isFruit(newFood)) {
       alert(newFood + ' is a fruit! Easter egg!');
   } else {
       addNewFoodItem(newFood, foodKey);
       window.localStorage.setItem(foodKey, newFood);
   }

  foodField.value = «»;
  return false;
});
```

Test it in Safari to make sure things work as expected:

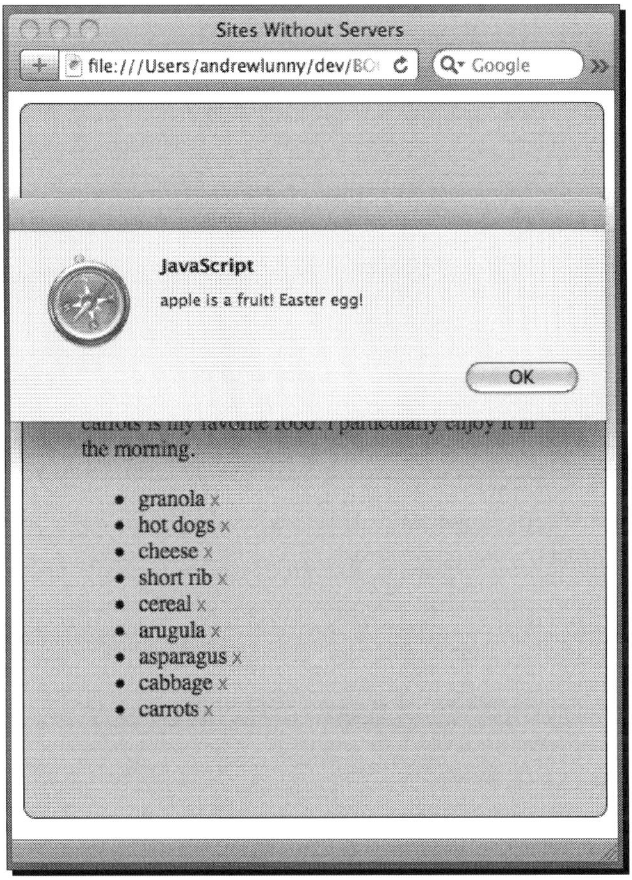

3. The next step is to include the video on the page. Ensure that you have the **phonegap-video.mp4** file in the same directory as the rest of the application—this is available with the code samples included with this book. Firstly, we'll write out the tag for the video element, to save as a JavaScript variable:

```
var videoTag = '<video id="sampleVideo" width="240" height="135"'
+
    'autobuffer src=»phonegap-video.mp4» controls></video>';
```

4. Most of that should look quite standard—it's very close to what the familiar `img` tag looks like. The most important part is the `controls` attribute—this ensures that the browser will render standard controls for the user to alter playback. If desired, you could forego this attribute and handle all of the controls manually with JavaScript.

5. Now for the actual display logic. We're going to insert the video into the DOM, play it right away, and set up a couple of event handlers. Using XUI, this is just a small amount of code. Firstly, we'll add a `div` to the top of our page, to contain the video:

```
<div id="main">
    <div id=»noVideo»></div>
   <h1>List Some Food</h1>
```

   Now here's the JavaScript function to show the video and start it playing:

```
function showAndPlayVideo() {
    if (!x$('#sampleVideo').length) {
        x$('#noVideo').inner(videoTag);
        x$('#sampleVideo')[0].play();
        x$('#sampleVideo').on('click', function () {
            this.pause();
        }).on('pause', function () {
            x$(this).remove();
        });
    }
}
```

6. If XUI can't find the `sampleVideo` element on the page, it inserts it into the `noVideo` div. Using XUI's array index syntax, we can access the DOM object itself and call `play` right away. We listen for the standard `click` event and call `pause` when it occurs (note that with XUI, `this` is bound to the DOM object for event handling). On the `pause` event, we remove the `sampleVideo` element from the DOM. Phew!

7. Let's tie all this together and expose our easter egg to our user. Go back to the form handler and replace the alert call with one to `showAndPlayVideo`:

```
if (isFruit(newFood)) {
    showAndPlayVideo();
} else {
```

Go back to Safari and check things out:

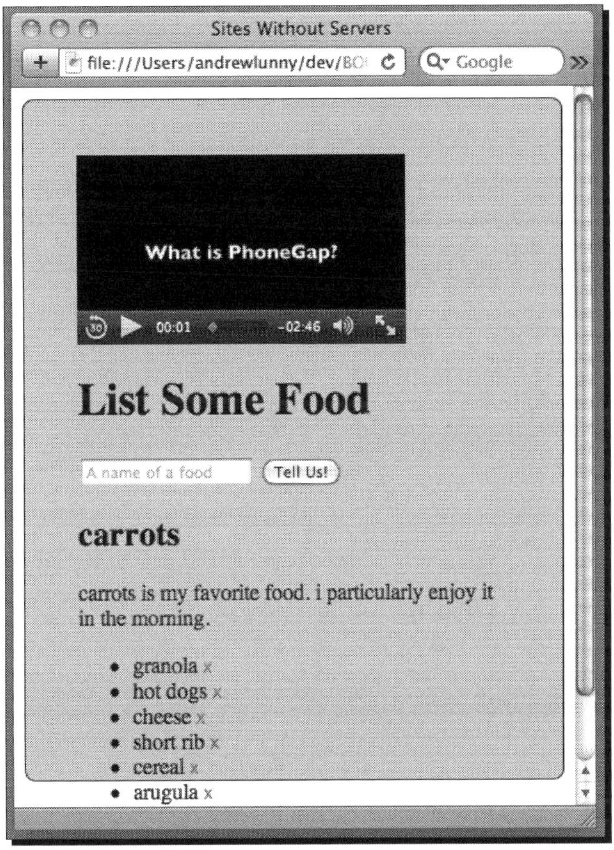

What just happened?

This was our first glimpse of the new media elements in HTML5—the interface for the `audio` element is basically the same as that for `video`. We can notice a couple of things right off the bat.

The video is just another DOM element on the page, so we can, theoretically at least, manipulate it much like any other DOM element—adjust the height and width, show and hide with CSS, bind handlers for the standard events, add and remove from the DOM.

Once we can grab the DOM object for the video through the JavaScript interface, some basic programmatic controls—`play` and `pause`—are very easy to use. One other useful property to be aware of is `currentTime`—the, yes, current time of the video, in terms of seconds elapsed. This is a writable property, so we could for instance, change our click handler to restart the video instead of removing it:

```
x$('#sampleVideo').on('click', function () {
    this.pause();
    this.currentTime = 0;
    this.play();
})
```

The `play`, `pause`, and `currentTime` controls make it very easy to implement your own controls for a video, rather than using the default ones provided by the browser.

Media events and attributes

We've seen that a `pause` event is fired by the video when playback has stopped, and, as you may expect, there is also a `play` event. In the event of playing a video from a network source, as opposed to the local file system, you would probably want to use the `canplay` and `canplaythrough` events also. Both indicate how much of the video's data is available—when `canplay` fires, there is enough to begin playback; when `canplaythrough` fires, there is enough to play the entire video.

We could tweak our above code to only call `play` when `canplay` is fired, rather than after the element is inserted in the DOM. In my experience, when dealing with a video from the local file system, it is safer to call play, since canplay may fire right away. Another option is to set the Boolean autoplay attribute on the video, like so:

```
<video src="myvideo.mp4" autoplay></video>
```

Your mileage may vary, particularly on different target platforms, so be aware that there are a few different ways to do things. Android in particular has had some problems with the `video` tag; for most success, be sure to explicitly call `play` through the JavaScript interface. A good reference on the subject of video on Android can be found here: `http://www.broken-links.com/2010/07/08/making-html5-video-work-on-android-phones/`.

Of the attributes in our sample tag, `autobuffer` ensures that the video starts downloading as soon as the browser encounters the tag, while `controls`, as mentioned above, displays the custom controls when the user activates them.

There are some other attributes that are not well supported on mobile devices, at the time of writing: `loop`, another Boolean attribute, loops the current video, while `poster` can be set to an image that is displayed until the first frame is loaded.

One final useful attribute is `webkit-playsinline`, which is just of use on iOS handheld (that is, non-iPad) devices. Assuming that the webview supports this property (which PhoneGap does), this attribute allows the video to be played in place, rather than switching to a fullscreen view. This can be very useful for the design of your application.

A final note on iOS video playback: there is a bug (which, again, may be fixed by the time you read this), which is that touch events that not fired on video elements. If your application prevents the user scrolling the viewport by cancelling all `touchmove` events, as is common with many PhoneGap applications, you will need to manually call `window.scrollTo` to reset the scrolling if things get misaligned. There's nothing worse than a user just seeing the right-hand side of your entire application.

The audio element

The media elements were designed in tandem to have similar APIs, so what you've read about the video tag should apply, for the most part, to the audio tag also.

The audio element has gotten less attention than its audiovisual cousin, and for that reason has arguably bigger support on the major mobile platforms than the video tag. In particular, many developers have found the buffering for audio elements to be subpar, which hinders the use of the audio element for sound effects, or immediate aural cues. The best advice is for careful and conscientious testing of audio in your application when you're using it, in order to avoid some of the pitfalls of unreliable support.

An interesting side note is the audio data API, which is beginning to be implemented on desktop browsers, although it's still quite far off for mobile developers. Using this API, JavaScript developers can have access to raw audio feeds, allowing for programmatically generated or manipulated audio right in the browser. This should allow for some fascinating new games and multimedia applications in the future; it's definitely something to look forward to.

Pop Quiz – Media elements

1. Which of the following is NOT possible using the HTML5 video API, with a 30 second long video clip?
    a. Looping over two seconds in the middle
    b. Playing the entire video backwards
    c. Starting playback 20 seconds in

2. What is the difference between the `canplay` and `canplaythrough` events?
    a. `canplay` fires continuously once the video starts buffering, while `canplaythrough` only fires once
    b. `canplay` fires when the video is loaded, while `canplaythrough` fires when the video is loaded and the user has pressed play
    c. `canplay` fires when some of the file is downloaded, while `canplaythrough` when the entire file is available
3. If a video is set to `autobuffer`, when does it start to download?
    a. When the user presses play
    b. Once the browser parses the appropriate `video` tag
    c. When the page loads

The canvas element

We've mentioned that one of the selling points of HTML5, in comparison with the hated Flash, is its support for dynamically creating graphics. There are two parts of the HTML5 specification that control dynamic graphics: the `canvas` element, for bitmap graphics, and inline **Scalable Vector Graphics**, **SVG**, for, well, vector graphics.

The two APIs are quite far apart in a number of ways: most notably, SVG graphics are made up of DOM elements, and can be manipulated easily using CSS and DOM events, a bit like the media elements in the previous section. The canvas, by contrast, is solely controlled by your own JavaScript; though basic touch and click events are conveyed by the canvas element, you have to figure out which of your objects on the canvas are being touched to fire any specific events.

Which should you use? For mobile applications, using PhoneGap or not, there isn't much of a choice. As of the 2.3 OS release, no Android device supports SVG in the native web view (a third-party browser, Firefox Mobile, does have support, but is not used by PhoneGap). Until Google implements SVG on Android, mobile web development has to go with canvas.

A final note: in previous chapters we alluded to the fact that CSS3 three-dimensional transforms are hardware accelerated on iOS devices. This is important because canvas animations are *not* hardware accelerated, either on iOS or anywhere else. Our example will deal with static graphics, so this will not be an issue, but for more intensive canvas applications you should keep this consideration in mind, and do as much testing on actual devices as possible.

Time for action: Dinner dashboard

We're going to set up a small dashboard next to the list of foods that will display circles to indicate how many foods are in the list—a red circle for five foods, a yellow circle for one:

1. The first thing we're going to do is add an empty canvas to our page. The canvas itself is just an HTML element, so we can style it through CSS but, to ensure things work as expected, we need to specify the width and height as attributes of the element. Here's where our tag fits in:

```
<div id="noVideo"></div>
<canvas id=»myCanvas» width=»50» height=»400»></canvas>
<h1>List Some Food</h1>
```

And here's how we'll style it with CSS:

```
#myCanvas {
    position: absolute;
    z-index: 0;
    left: 280px;
    opacity: 0.8;
    background: #ccf;
}
```

Take a quick look in Safari:

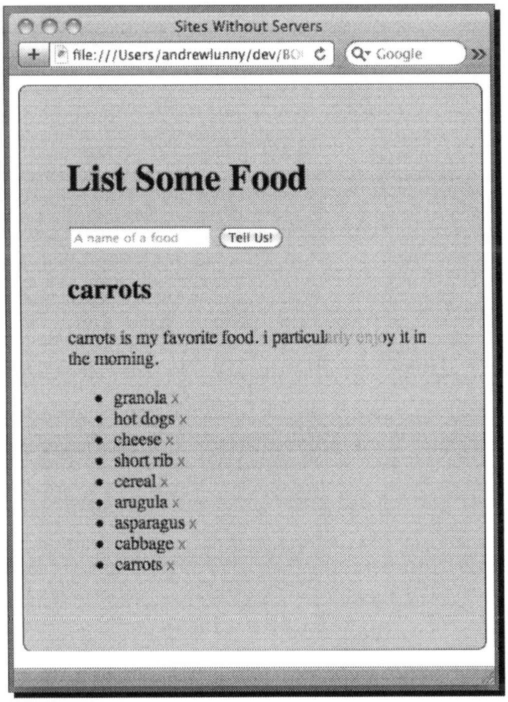

2. The first thing we'll do is draw a yellow circle on the canvas, by writing a JavaScript function to do so:

```
function drawCircle() {
    var canvas = x$('#myCanvas')[0];
    var ctx = canvas.getContext('2d');

    var x = 25, y = 35, rd = 10;

    ctx.beginPath();
    ctx.arc(x, y, rd, 0, Math.PI * 2, false);
    ctx.closePath();

    ctx.fillStyle = «yellow»;
    ctx.fill();
}
```

3. There's a lot going on in there, which we will cover shortly. For now, add a call to `drawCircle` to the `DOMContentLoaded` event handler:

```
document.addEventListener("DOMContentLoaded", function () {
    drawCircle();
    ...
```

And check again in Safari:

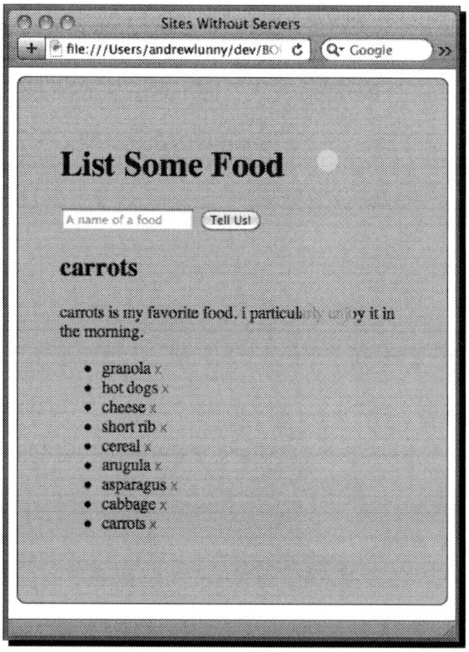

4. That's a good start. Now we need to draw a dashboard on our canvas, so let's start writing out `drawDashboard`. It should take the quantity of foods we have, draw a red circle for each five foods, and then a yellow circle for each of the remainder. Here is what our function should look like:

```
function drawDashboard(length) {
    // double-tilde forces integer division in JavaScript
    var reds = ~~(length / 5);
    var yellows = length % 5;
    var count = 0, i;

    for (i=0; i < reds; i++) {
        drawCircle('red', count++);
    }

    for (i=0; i < yellows; i++) {
        drawCircle('yellow', count++);
    }
}
```

5. The next step is to modify our `drawCircle` to do the right things, given these parameters. It should take the color, and also the offset for each circle:

```
function drawCircle(color, offset) {
    var canvas = x$('#myCanvas')[0];
    var ctx = canvas.getContext('2d');

    var x = 25, y = 35 + (30 * offset), rd = 10;

    ctx.beginPath();
    ctx.arc(x, y, rd, 0, Math.PI * 2, false);
    ctx.closePath();

    ctx.fillStyle = color;
    ctx.fill();
}
```

6. And let's change our `DOMContentLoaded` call to use `drawDashboard`:

```
document.addEventListener("DOMContentLoaded", function () {
    drawDashboard(13);
```

Back to Safari to check all is well:

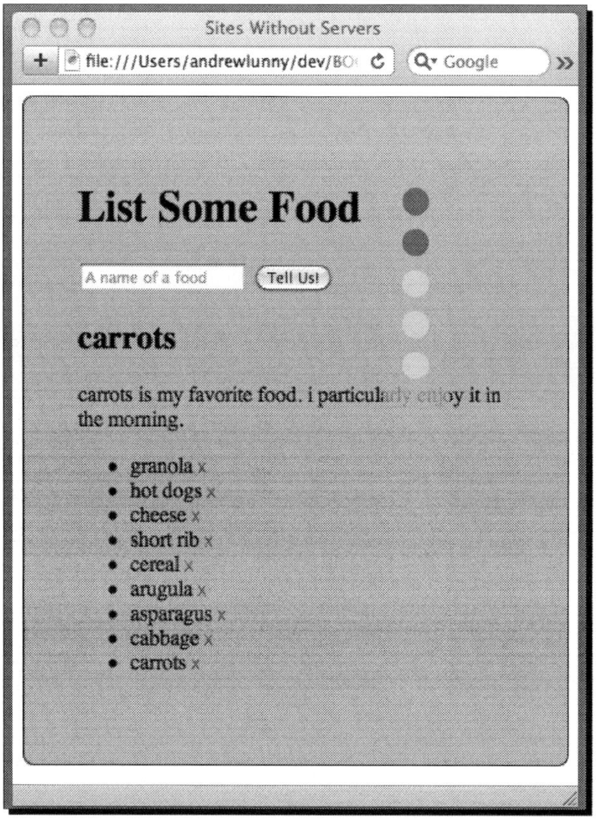

7. Now to make that correspond to the actual display, remove the `drawDashboard(13)` call and add the following after the list is populated:

```
x$('li').css(defaultStyle);
drawDashboard(x$('li').length);
```

8. That sets the initial state of the dashboard just fine, but we need it to be responsive, based on when the number of items in the list changes. Each time that happens, we're going to redraw the dashboard.

9. The number of items in the list either when one is added or one is removed: either way, we're going to be calling `drawDashboard` again. A food is added through the submit handler on the form, so we can call the function there:

```
if (isFruit(newFood)) {
   showAndPlayVideo();
} else {
    addNewFoodItem(newFood, foodKey);
    drawDashboard(x$('li').length);
    window.localStorage.setItem(foodKey, newFood);
}
A food is removed when a deleter span is clicked on, so we can add
a call there also:
} else if (x$(evt.target).hasClass('deleter')) {
    var listIem = x$(evt.target.parentNode);
    var storageKey = listIem[0].id;

    listIem.remove();
    window.localStorage.removeItem(storageKey);
    drawDashboard(x$('li').length);
}
```

10. This is almost there but if we look in Safari, it's not quite working—the dashboard is just overwriting itself, not clearing correctly each time a new value is being written.

11. Clearing a canvas element is quite easy—you just need to write to the `width` property, or the `height` property. Let's add that to `drawDashboard`:

```
function drawDashboard(length) {
    // double-tilde forces integer division in JavaScript
    var reds = ~~(length / 5);
    var yellows = length % 5;
    var count = 0, i;

    var canvas = x$('#myCanvas')[0];
    canvas.width = canvas.width;

    for (i=0; i < reds; i++) {
        drawCircle('red', count++);
    }

    for (i=0; i < yellows; i++) {
        drawCircle('yellow', count++);
    }
}
```

Now if we look at Safari, we have a responsive and immediate dashboard at hand:

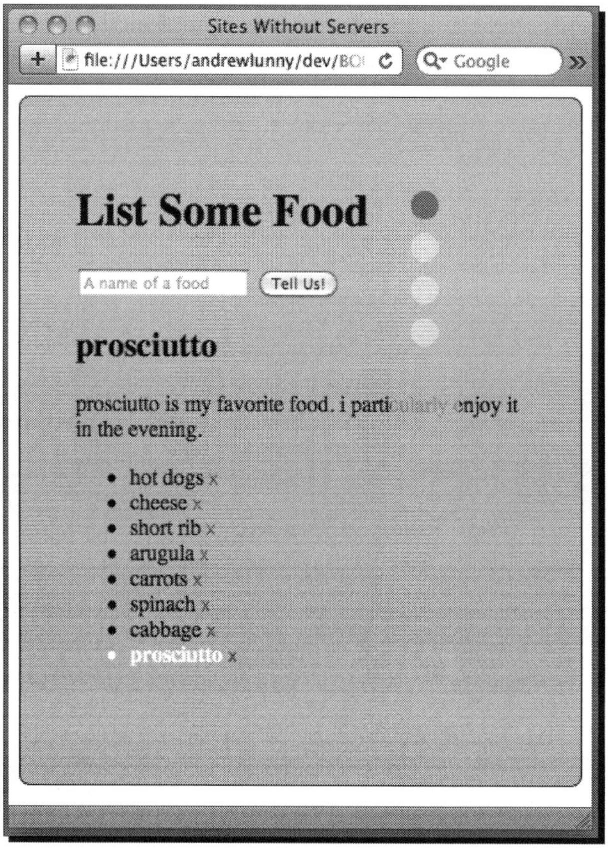

What just happened?

We took a look at the baseline functionality of the canvas element: grabbing a context and drawing a shape directly to it. A circle (a special kind of arc) is very nice, but we could also have drawn line paths or rectangles easily, using the canvas API. For this kind of drawing, it's very easy to get started with the canvas, and you can play around in Web Inspector until you find something that you like.

The canvas API

As, essentially, the major HTML5 competitor to the Flash juggernaut, the canvas has far too broad an API to cover in full in this chapter. Indeed, there are entire books written about the things, and an increasing number of libraries, tools, and even IDEs emerging to create animations using the canvas. But a few points are worth touching on.

Look back at the first lines of our drawCircle function:

```
function drawCircle(color, offset) {
    var canvas = x$('#myCanvas')[0];
    var ctx = canvas.getContext('2d');
```

The third line there, with the call to getContext('2d'), is an essential part of any of your canvas code. The canvas object exposes a context property upon which the code can directly draw. There are functions available for 2D shapes: bookend with beginPath and closePath, use moveTo and lineTo for linear paths and arc for curves.

You can also draw images and text directly to the canvas: drawImage for images and fillText for text. drawImage is passed an image object directly, not a path; unless the image is already on the DOM, you'll need to use the unfamiliar new Image() idiom to get it onto the canvas:

```
var img= new Image();
img.src = "img.png";
img.onload = function() {
    ctx.drawImage(img, 0, 0);
};
```

This new Image() technique is not used often for adding images to the DOM, but is very useful with the canvas, which has a pure JavaScript API.

Why draw images to the canvas? Well for one thing, you could draw over them, skew them, and rotate them (although much of that is possible with CSS3). What's nice about the canvas is that it provides a function called getImageData, which allows you to pull the pixel data out of the canvas element, in order to transform it, or to save as a PNG file. We haven't yet seen the full extent of what people can do with this technique, but there are already some very interesting libraries coming out, such as Dave Shea's **PaintbrushJS**, available at https://github.com/mezzoblue/PaintbrushJS, which applies Photoshop-style filters to images in real-time, on the client side.

A note on performance

The code in the prior example was designed for clarity of behavior, not for performance, so there are one or two caveats to take note of.

First, if at all possible, only clear the section of the canvas that you have to redraw. A clearRect function is available on the 2D context that does just this job—if you know which area of the canvas is "dirty", it's far more performant to just clear that section.

Secondly, try to minimize the amount of strokes and fills you perform. This example, with some discrete circles, is not hugely affected, but if you are drawing any especially complex line segments to the canvas, performance suffers greatly unless you're sparing with the calls to `stroke`.

Have a go hero

Can you make the dashboard a little more interactive? How would you handle clicks or touches on the dashboard—is there an easy way to identify which part of the canvas was interacted with, and whether a yellow circle or a red one was touched?

If you're feeling especially heroic, you may want to try performing some animation on the canvas: can you touch a red circle and have it expand into five yellow ones?

If you need further reference on the `canvas` element, check out, again, Mark Pilgrim's Dive Into HTML5, at `http://diveintohtml5.org/canvas.html`. There's also an excellent tutorial available from the Mozilla Developer Network, `https://developer.mozilla.org/en/Canvas_tutorial`.

What else is in HTML5?

One of the most popular uses of HTML5 is for location services: the `navigator.geolocation` object gives a reading of where the user's current location is. Since PhoneGap handles this a little different to vanilla web browsers, we'll cover geolocation as a PhoneGap API.

In a similar boat is the application cache, or **cache manifest**. This allows web developers to specify a list of static files that the browser should attempt to cache for future requests. With PhoneGap applications running from the local file system, this is not applicable (and non-existent on the `file://` protocol, at any rate).

Two of the most exciting new features of HTML5 are **Web Workers** and **Web Sockets**, neither of which is well-represented on mobile devices at the moment. Web Workers, a lightweight concurrent processing approach, are a great idea, but there's no way to simulate them on current devices in a manner that doesn't degrade performance (Web Workers use threads to achieve concurrency, and simulating the API in a single-threaded JavaScript runtime would not provide true concurrency).

Web Sockets, persistent connections between a browser and a web server, are arguably more interesting from a server perspective than a client one. At any rate, the specifications are under heavy revision, and not stable enough for adoption on mobile platforms just yet.

Summary

In this chapter, we have seen the following:

- Mobile JavaScript libraries, such as XUI, can greatly increase the efficiency and expressiveness of the code that we write. In most cases, it's silly to attempt a new application without a solid library in your corner.
- The new HTML5 media elements provide an intuitive yet powerful API for improving the responsiveness and depth of interaction for your application.
- The HTML5 canvas element makes it very easy to create cross-platform, dynamically generated graphics, using the JavaScript skills that you are already learning.

In the next chapter, we will extend the skills we've learned about HTML5 with the help of its little brother, little CSS3.

6

CSS3: Transitions, Transforms, and Animation

Web developers will be long familiar with Cascading Style Sheets: CSS is the standard way to style web applications and web pages. As we have seen in earlier chapters, a good grasp of CSS is essential to design PhoneGap applications that look beautiful on their target devices. Whether you're working on a richly interactive canvas-based game or a straightword buttons and list application, CSS3 features will allow your application to shine that much more.

In this chapter, we'll see some essential spots where CSS3 techniques and libraries can be used, and also some cool effects we can add to give our applications a bit more shine. In particular, we will:

- Expand on our use of transitions to easily add native-style animations to paging applications
- Explore one of the ever-popular scrolling libraries to implement one of the holy grails of mobile web design: a fixed position toolbar!
- Get a taste of CSS3 explicit animations, and an idea of their power and flexibility

Translate with transitions

For a PhoneGap developer, CSS3 transitions—particular the `translate` and `translate3d` transforms—are some of the most important and widely used features of modern browsers. It's pretty easy to see why:

- Your old-fashioned web application runs across multiple web pages. To change from one state of the application to another, the old page is cleared from memory and a new one is loaded—usually the screen goes white for a moment.

- Your new-fangled PhoneGap application runs from a single web-page (other than in exceptional circumstances). To change from one state of the application to another, well, let's see...

We could, of course, just hide the old data and display some new things. That's a perfectly viable solution when you're developing your application initially, and you just want to ensure that all of the logic works correctly, but for a finished application you typically want a lot more polish.

That's where transitions and transforms are most often used—to give a nice veneer to the business of moving your application between states. As with most native mobile applications, a PhoneGap application gains greatly in perceived responsiveness from a little animation—a smooth slide from right-to-left looks much better than a flash of white space.

We're going to go back to our much loved Food List application and up its sheen a little. Instead of using Web-style alert boxes, we're going to use CSS3 transforms to display new tweets.

Time for action – The modal tweet view

We're going to make a couple of brief changes to the code from the previous chapter, in order to simplify what's going on, and make a quick detour into debugging territory:

1. Firstly, we want a clearer divide between our HTML markup, our JavaScript code, and our CSS rules. Cut out the contents of the `style` tag (there should just be one) and paste it into a new file called `style.css`. Do the same, into a file called `food-list.js`, for the contents of the application's two `script` tags.

   We then need to add a couple of tags to our page to reference these external files. Add a `link` tag to the head of the document:

   ```
   <head>
       <title>Sites Without Servers</title>
       <link rel=»stylesheet» href=»style.css» />
   </head>
   ```

   And a `script` tag to the end of the document (make sure our two other `script` tags, referencing `xui.js` and `mustache.js`, are also moved down):

   ```
       </body>
       <script src=»xui.js»></script>
       <script src=»mustache.js»></script>
       <script src=»food-list.js»></script>
   </html>
   ```

Moving the `script` tags to the bottom of the page helps a little with load times, especially on public-facing websites—it doesn't make a huge difference for PhoneGap applications, but it's a good habit to get into.

2. Firstly, let's test out that everything works as before—in particular, we want to edit the behaviour when a tweet is retrieved and displayed to the user. Open `index.html` in Safari and try clicking on one of the foods, in order to display the latest tweet.

   If you've been following along with the last chapter, you'll see that all is not well; in fact, nothing at all is being displayed. Let's open up the Web Inspector and see what error we're getting:

   TypeError: Result of expression 'latestTweet' [undefined] is not an object.

3. We know `latestTweet` should be set based on the request we make to the Twitter search API—evidently, that's not happening. Luckily, Web Inspector includes a **Resources** panel—this allows us to look at all of the requests our application makes, and all of the responses that it gets back. Open the Resources panel and look at our request to Twitter:

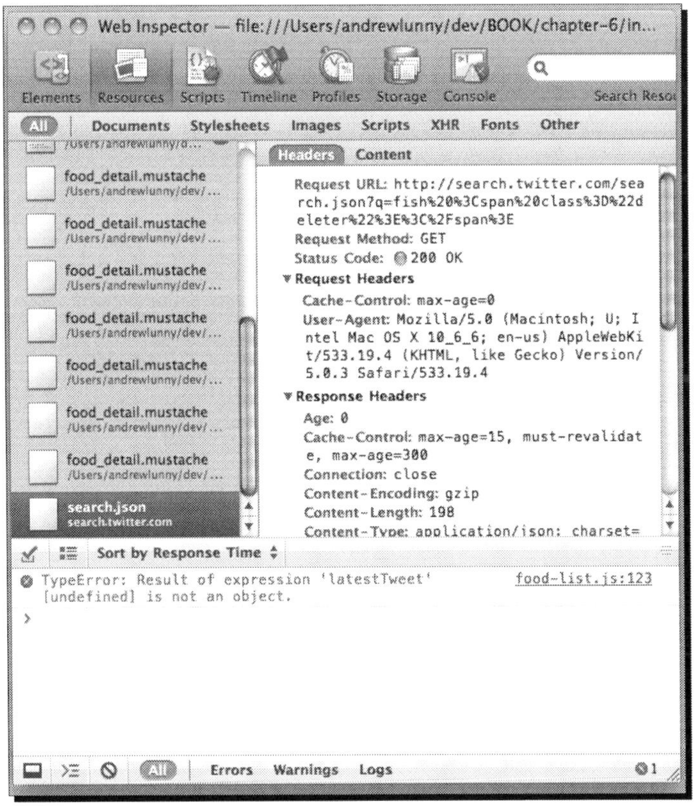

4. Check out the **Request URL**—the search query that we're giving to Twitter:

```
http://search.twitter.com/search.json?q=fish%20%3Cspan%20class%3D%22deleter%22%3E%3C%2Fspan%3E
```

5. It looks like we're sending the markup of our `deleter` button along with the name of the food—no wonder we're not getting any results! Luckily, we can fix this quite quickly—we just need to use the element's innerText property rather than innerHTML. Find the relevant line in `food-list.js`, and it's an easy change:

```
x$(document).on("click", function (evt) {
    if (evt.target.tagName == «LI») {
        var foodSubject = evt.target.innerText ||
        evt.target.textContent;
        var foodSearch = encodeURIComponent(foodSubject);
        var twitterUrl =
        «http://search.twitter.com/search.json?q=» + foodSearch;
```

6. That should solve it—let's fire up Safari again to make sure our problem is solved:

Phew! Now back to styling.

7. If you remember, we want to set up a modal view window to display the latest tweet. This modal window will take up all of the view, and display the tweet as white text on a black background. The first thing to do is set up this view, with some dummy data. Let's add the markup to `index.html`, just below the closing tag of the div with ID `main`:

```
<div id="modal">
    <h1>Latest Tweet about «muffins»</h1>
    <blockquote>Writing a book is hard, but at least I have some
     muffins</blockquote>
    <p>by <strong>alunny</strong></p>
</div>
```

8. Now we need to style that view to be in the right place, and look correct. We want to inherit the basic styles of `div#main`, so we can change that selector to `div#main, #modal`. Next, let's define the style of the `#modal` div itself:

```
#modal {
    background-color: rgba(0,0,0,0.8);
    border-radius: 0;
    color: white;
    font-family: helvetica, sans-serif;
    left: 0;
    position: absolute;
    top: 0;
}
```

Note the background-color—we're using RGBA color, which allows us to specify an opacity (the fourth parameter). The opacity is between 0 (fully transparent) and 1 (fully opaque). At any rate, let's load it in Safari and see how it looks:

9. Let's get that to be dynamically populated. Firstly, we'll remove the dummy text, and make each section easier to find:

```
<div id="modal">
    <h1>Latest Tweet about «<span id=»search_term»></span>»</h1>
    <blockquote id=»tweet»></blockquote>
    <p>by <strong id=»author»></strong></p>
</div>
```

We're also going to modify the CSS to be hidden by default:

```
#modal {
    background-color: rgba(0,0,0,0.8);
    border-radius: 0;
    color: white;
    display: none;
    font-family: helvetica, sans-serif;
    left: 0;
    position: absolute;
    top: 0;
}
```

11. Now let's modify the code that displays the tweet alert, changing its anonymous callback function to call a new function, `showTweetModal`:

```
getXHR(twitterUrl, function (response) {
    var latestTweet = getLatestResult(response);
    showTweetModal(latestTweet, foodSubject);
})
```

Then write the `showTweetModal` function—with XUI and our new markup, it's an easy function to get right:

```
function showTweetModal(tweetObject, searchTerm) {
    x$(«#modal #search_term»).inner(searchTerm);
    x$(«#modal #tweet»).inner(tweetObject.text);
    x$(«#modal #author»).inner(tweetObject.from_user);

    x$(«#modal»).css({«display»:»block»});
}
```

Give it a shot now—reload Safari and click on a tweet:

12. Looks delicious! To keep things simple, we're going to hide the modal any time somebody clicks the black space. We'll add another clause to our `click` event delgation (the call to `addEventListener`):

```
} else if (evt.target.id == "modal") {
    hideTweetModal();
}
```

Then write a simple `hideTweetModal` function:

```
function hideTweetModal() {
    x$(«#modal»).css({«display»:»none»});
}
```

13. Okay, so the basic functionality is in there—we now need to get the transitions in place. Let's replace `display: none` in our CSS `#modal` rules for something more helpful:

```
#modal {
    background-color: rgba(0,0,0,0.8);
    border-radius: 0;
    color: white;
    -webkit-transition-duration: 800ms;
    -webkit-transform: translate3d(0,600px,0);
    font-family: helvetica, sans-serif;
    left: 0;
    position: absolute;
    top: 0;
}
```

When you reload the page in Safari, you'll notice it looks much the same—because we've positioned the modal view 600 pixel lower, beneath the bottom of the window, it appears to be hidden.

14. How do we rewrite `showTweetModal` and `hideTweetModal` to accommodate for our new style? It's an easy change to make—as with our initial example, `showTweetModal` changes the default style and `hideTweetModal` restores it:

```
function showTweetModal(tweetObject, searchTerm) {
    x$(«#modal #search_term»).inner(searchTerm);
    x$(«#modal #tweet»).inner(tweetObject.text);
    x$(«#modal #author»).inner(tweetObject.from_user);

    x$(«#modal»).css({«-webkit-transform»:»translate3d(0,0,0)»});
}

function hideTweetModal() {
    x$(«#modal»).css({«-webkit-transform»:»translate3d(0,600
      px,0)»});
}
```

Check things out in Safari—the following screenshot is from mid-transition. Watch that amazing, smooth sliding action!

What just happened?

Unlike our simple example in an earlier chapter, here we got to implement a practical use for CSS3 transitions. In this case, we're mimicking the standard iOS modal display: a quick display that slides up from the bottom of the screen and, when dismissed, slides back down. These, in the present author's opinion, work well on mobile applications for displaying brief information quickly to the user, perhaps with a couple of options for interacting with it. Alert boxes, whether using the built-in JavaScript `alert` function or PhoneGap's `navigator.notification.alert` function, are better suited for simple notifications: "You have a new email message", or whatnot.

In both these examples we've used the `translate` (or rather, `translate3d`) CSS function for our transitions. These are the simplest transformation functions to apply—each of them moves the element and its children absolutely by the values specified, along the x, y, and z dimensions. `Translate3d` is prefered to `translate` as 3D transitions are hardware-accelerated on iOS devices (and similar acceleration has been promised on other platforms, including Android). In practice, 2D translations don't look too bad on Apple devices, while they can both be jerky on other platforms. Your best bet is to test on the devices that you're targeting, and modify your animations appropriately.

Timing functions

Our example used CSS3 transforms as a pretty blunt instrument—we told the CSS where our modal display should go, and how long it should take to get there, but nothing further. However, there are other properties available in the CSS3 specifications that give finer granularity over the behavior of these animations.

One that is very useful is `-webkit-transition-timing-function`, which allows you to specify how the transition should accelerate or decelerate as it gets from a to b. We can either give a built-in function to the transition, or use the `cubic-bezier` function to specify the speed curve of the transition. Since that math is beyond a humble JavaScript developer like myself, we're going to use the built-in functions to spice up our transitions a little:

```
function showTweetModal(tweetObject, searchTerm) {
    x$("#modal #search_term").inner(searchTerm);
    x$("#modal #tweet").inner(tweetObject.text);
    x$("#modal #author").inner(tweetObject.from_user);

    x$("#modal").css({
        "-webkit-transform":"translate3d(0,0,0)",
        "-webkit-transition-timing-function":"ease-in"
    });
}

function hideTweetModal() {
    x$("#modal").css({
        "-webkit-transform":"translate3d(0,600px,0)",
        "-webkit-transition-timing-function":"ease-out"
    });
}
```

This small change gives a bit of weight and momentum to the transition; feel free to tweak the timing function and transition duration properties a little more to get the movement exactly how you would like it.

Other transformations

There are many other transformation functions available in newer web browsers: `scale`, `rotate`, `skew`, `perspective`, and, most generally, `matrix`. For your average bog-standard lists and buttons mobile application, the `translate` and `translate3d` functions will take you pretty far; most native applications do little more graphically than slide views into and out of the viewport.

There are some exceptions though: scrolling libraries, which we will cover in the next section, depend heavily on `matrix` transformations and timing functions to present a smooth appearance of scrolling, as well as momentum-based scrolling, where the speed of the scrolling is based on the speed of the touches on the screen.

The easiest way to get started with the more complex CSS transformations is to look at a library that wraps them: the easiest one of those is probably **jQTouch**, `http://jqtouch.com/`, created by David Kaneda and now maintained by Jonathan Stark. jQTouch allows easy usage of complex CSS transform effects, such as cube rotation and scaling views in or out. For a larger, but equally impressive, solution, check out the **Sencha Touch** framework, which you can find at `http://www.sencha.com/products/touch/`: Sencha Touch can afford a great visual sophistication to your mobile applications, although you'll have to write your application from scratch to take advantage of the Sencha way of doing things.

Have a go hero: CSS transforms

Our example had a single modal view, with all of the markup in place, where we just had to populate a couple of text nodes with the appropriate content.

Can you use our transitions for a more complex application? If your application had a list of different views that would appear, each populated with different content, how would that change your approach to CSS3 transforms? How would you combine these transforms with a templating library, such as Mustache? Try combining these different techniques. If you can keep your code nicely decoupled, and ensure that the animation logic is kept separate from the rendering logic, you should be able to get some interesting things going.

Scrolling

The most frequent usage of CSS3 transforms on mobile devices is for implementing fixed positioning—or, more accurately, for enabling scrolling panels within a web view, and thus fixing the position of elements outside of the scrolling panel.

Since the debut of the iPhone, if not earlier, a few simple design patterns have emerged for mobile user interfaces. The most prevalent, on a portrait-oriented page, is a fixed title bar at the top of the page, a scrollable content panel in the middle section, and a fixed toolbar at the bottom. With a few variations, this is common on most other platforms as well; on Android devices, which have a menu button, the bottom toolbar is rarer, but you will still often see a fixed section at the top, with or without further options present.

Why is this not possible on mobile browsers out of the box? For the answer, let's explore the exciting world of viewports.

Viewports: Visual and otherwise

It's important to understand how mobile browsers render pages, in order to see why the CSS `position: fixed` rule fails to work as expected. A page is first rendered at roughly the width of a desktop browser; in the iPhone's case, the default width is 980 pixel. We can change this, as we have seen, by using the following meta tag:

```
<meta name="viewport" content="width=device-width,initial-scale=1.0"
/>
```

This is probably the best viewport setting to use for a PhoneGap application: allowing the user to scale the entire view makes your application look more like a standard web page than a native application. The intial scale the page is rendered at is known as the **layout viewport**; this distinguishes it from the **visual viewport**, which contains what the user sees (again, in most cases, when you're developing a PhoneGap application you want the two to be equivalent).

When the user scrolls on a mobile browser, the visual viewport is moved over the page, rather than the page moving itself. Pertinently, any elements that have been set to `position: fixed` are fixed in their position on the layout viewport, not on the visual viewport. Practicially, `position: fixed` is the same as `position: absolute`.

Since this is such a common problem, there are a number of libraries and approaches to implement something like `position: fixed`. It turns out it's easier to fake the scrollable part of one of these layouts than to actually fix the position of particular elements.

iScroll

Large user-interface frameworks, such as Sencha Touch, include their own scrolling code for this purpose, but there are a number of single-purpose libraries that have just this functionality. One of which, **GloveBox**, found at `http://purplecabbage.github.com/GloveBox/`, is developed by Jesse Macfadyen, one of the core PhoneGap iOS developers; it's particularly useful for layouts with multiple scrollable panels, which are well-suited for iPad applications. We're going to use **iScroll**, which is the most popular one of these libraries, developed by Matteo Spinelli. iScroll is available at `http://cubiq.org/iscroll`. We'll be using iScroll later in our next tutorial, so you should download it now.

All of these libraries do broadly the same thing: attach to the `touchmove` event on a particular element, cancel the default behavior (native scrolling), make a calculation based on how far the touch has moved, and move the content within the panel appropriately. The technique probably is best known from Google's mobile Gmail web page; if you're interested in implementing it yourself, there's an excellent article from Google's developers available here: `http://code.google.com/mobile/articles/webapp_fixed_ui.html`.

> One last note: all of these libraries depend on touch events to perform correctly, so we'll have to view them on either a mobile device or an emulator.

Time for action – Scrolling list of food

As hinted in the above note, we'll need to get a mobile Webkit touch-enabled device, or at least an emulator, up and running to play with iScroll. Start a PhoneGap project for iPhone, Android, or BlackBerry 6+, and copy the food list code we've been working on into the www directory—make sure you can load up the app on the device or emulator before going any further.

Also, ensure that you have Internet access on your emulator—for Android, you may need to change some settings, as outlined in this blog post: `http://www.paulmccrodden.com/2010/12/android-emulator-internet-access/`

1. First, make sure the correct viewport meta tag is on the page. The head of `index.html` should look like this:

```
<head>
    <title>Sites Without Servers</title>
    <link rel=»stylesheet» href=»style.css» />
    <meta name=»viewport» content=»width=device-width,
     initial-scale=1.0» />
</head>
```

2. Confirm that everything looks as expected—I'm using the iPhone Simulator here, but you'll want to check using whichever platform you have at hand:

3. It looks alright, but we know if keep adding food to the list, eventually it's going to go off the screen—that's why we're using iScroll. The first thing to do is add iScroll to our page:

```
<script src="xui.js"></script>
<script src=»mustache.js»></script>
<script src=»iscroll.js»></script>
<script src="food-list.js"></script>
```

4. We need to make sure the browser's normal `touchmove` event doesn't interfere with iScroll, which is going to override that event. Add the following code to `food-list.js`:

```
document.addEventListener('touchmove', function (e) {
    e.preventDefault();
});
```

Re-install the application on your device or emulator to confirm that regular scrolling is disabled (note that this disabled touch scrolling on your emulator; this will be when you hold the mouse down and drag the screen up or down).

5. Edit the mark-up on our page, adding a wrapper element to our list of foods:

```
<div id="wrapper">
    <ul id=»foodList»>
    </ul>
</div>
```

6. Now define the style rules for the `wrapper` element:

```
#wrapper {
    position: relative;
    z-index: 1;
    width: auto;
    height: 150px;
    overflow: scroll;
}
```

The most important properties here are the `position`, `z-index`, and `height`. You'll need to specify an absolute height for iScroll to work correctly.

Here's how the page looks now; note that the list of foods is clipped slightly:

7. Now we need to initialize iScroll—add the iScroll initialization line of code to the `DOMContentLoaded` **event handler in** `food-list.js`:

```
document.addEventListener("DOMContentLoaded", function () {
...
var scroller = new iScroll('foodList');
}
```

8. Open the application up once more to verify that everything works well:

What just happened?

We saw how easy it is to integrate iScroll in our application—a couple of lines of JavaScript, a few CSS rules, and a single extra element to add to our mark-up.

The ease of integration is one of the reasons iScroll is so useful as a tool for a PhoneGap developer. As we have seen, we can take a pre-existing layout and modify it to include this kind of scrolling with minimal effort. iScroll does a couple of interesting things as well as the scrolling—it adds the vertical and horizontal scrollbars to the wrapper element, and can monitor the DOM to refresh itself on any changes.

Since the baseline behavior is dealt with so well in iScroll, it's easy to add extra gloss on top of it; adding a gradient mask, for instance, can be done with CSS, in the same manner that you would do for any other element on the page. Things can get a little tricky if your DOM inside the scrollable panel is changing a lot, but a quick call of the `refresh` method on your new iScroll object should fix things up.

Other approaches

If iScroll, and the kinds of layout it enables, are so useful, why isn't this behavior baked into browsers by default?

The answer involves a lot of hand-waving, a reference back to the viewport section, and a frustrated sigh. Essentially, it's very performance intensive to do this in a browser, especially on the limited resources of a mobile device. Something like `position: fixed`, we imagine, will make it on to these devices at some point, but until then we have to do it at the application level, which involves a lot of messy computation that gets in the way of the fun stuff.

The alternative to the iScroll approach—faking scrolling—is to fake the fixed position element. Instead of having an element fixed at a certain position, the element is positioned absolutely, and then repositioned as the user's viewport is modified. This has the neat property of letting the browser take control of the scrolling, instead of doing it in JavaScript, and can be less processor-intensive. The popular **jQuery Mobile** library uses this approach for its fixed toolbar elements.

The downside to the repositioning approach: it doesn't look like a native application. Whether this is an acceptable trade-off for your users depends on the application itself; both approaches are certainly worth a shot.

Pop quiz – Scrolling

1. Why doesn't `position: fixed` work on many mobile browsers?
   a. With limited screen space, all the available space has to contain fresh data
   b. There's a mismatch between how pages are rendered on the visual viewport and the layout viewport
   c. Apple doesn't care about web developers
2. What does iScroll use CSS3 transforms for?
   a. To arrange the content element within the wrapper element
   b. To calculate how far the user has moved her finger
   c. To reproduce the momentum effects of touch scrolling

3. Can I forego using a scrolling library, and just use the web view's scrolling facilities?
    a. Yes: with a carefully thought-out design, you can easily manage the complexity of most mobile applications without resorting to JavaScript based scrolling
    b. No: the standard mobile design patterns exist for a reason, and users are trained to expect native scrolling in familiar layouts
    c. All of the above

Explicit animations

We looked at CSS3 transforms earlier in this chapter, as an example of simple, implicit animations that can be specified quickly. CSS3 also provides a syntax for defining animations in more detail—defining individual keyframes of an animation, and the relationships between different keyframes. Although the syntax for defining such animations is a lot more verbose than that for transforms, the flexibility and power of the resulting effects make these animations a superb tool.

Essentially, we define our own animation function, and can then apply it in a similar fashion to the built-in transformation functions, such as `translate` or `rotate`. For performing complex manipulations of DOM elements, CSS animations are a great solution.

Time for action – Animating our headline

I'm going to switch back to Safari for this section, in order to iterate faster on development. If you want to stick with your mobile device or emulator, be my guest.

1. We're going to animate the main header on the page, so let's set up it's starting state to begin with. Let's first change the markup slightly:

```
<h1 id="pageTitle">Food List</h1>
```

2. Then style this element a little:

```
h1#pageTitle {
    -webkit-border-radius: 9px;
    border: black 1px solid;
    padding: 6px 12px;
    background-color: black;
    color: white;
    width: 175px;
}
```

Check it out in Safari—the title should stand out a bit more now:

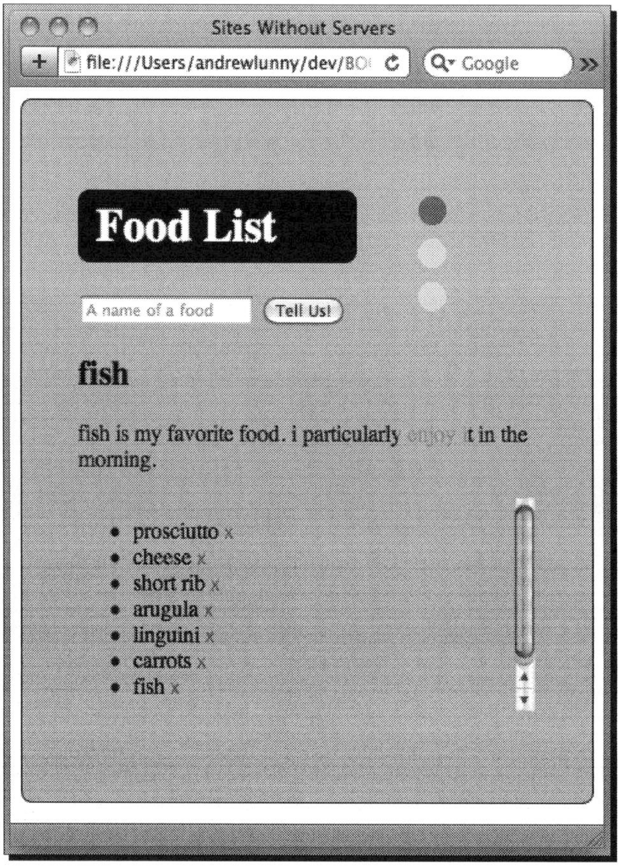

3. That title looks okay, but I bet we can make it a lot more garish. We're going to define a CSS animation called throbbing that will do just that. Firstly, we need to define the animation itself, as a set of keyframes—here's what that looks like:

```
@-webkit-keyframes throbbing {
    0% {
        background-color: black;
        color: white;
        -webkit-transform: scale(1.0) translate(0,0);
    }
    20% {
```

```
        -webkit-transform: scale(1.3) translate(40px,0);
    }
    50% {
        background-color: white;
        color: black;
        -webkit-transform: scale(1.2) translate(40px,-40px);
    }
    80% {
        -webkit-transform: scale(1.3) translate(0,-40px);
    }
    100% {
        background-color: black;
        color: white;
        -webkit-transform: scale(1.0) translate(0,0);
    }
}
```

See if you can figure out what this animation will look like—you can see we're adopting some of the transforms we've already seen, and also switching colors.

4. Now apply the animation to the `h1#pageTitle` element:

```
h1#pageTitle {
    ...

    -webkit-animation-name: throbbing;
    -webkit-animation-duration: 6s;
    -webkit-animation-direction: alternate;
    -webkit-animation-timing-function: ease-in-out;
    -webkit-animation-iteration-count: infinite;
}
```

5. The `duration` and `timing-function` properties map to those that we saw for applying built-in translations. Now reload the page in Safari to gaze on our fresh monstrosity:

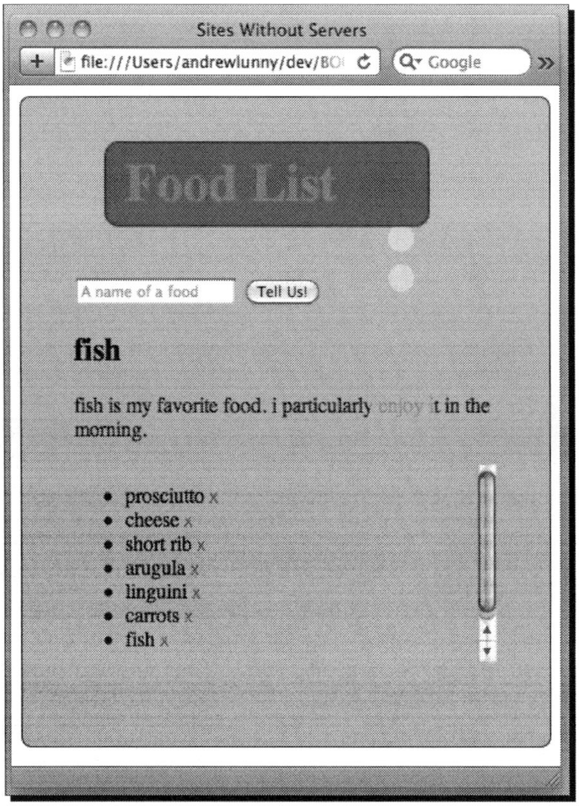

Who says you need Flash for ugly web pages!

What just happened?

We saw the syntax for user-defined CSS3 animations—it's a slight departure from what you may have been expecting from CSS, what with something like a function being defined and assigned to something like a variable.

It's a welcome change, in our opinion—far better than having to assign each keyframe as a separate class, for instance, and use JavaScript to figure out the timing function and apply each class as appropriate.

Here's a simplified view of the syntax we used to defined our throbbing animation:

```
@-webkit-keyframes animation-name {
    0% {
        /* original state */
    }
    /* at some point in the middle */ {
        /* properties to animate */
    }
    100% {
        /* final state */
    }
}
```

Easy as punch, right? We can add as many blocks between 0 and 100 as we wish (well, any more than 99 might be pushing it) and choose whichever properties we wish to be animated. Of course, some properites make more sense to animate than other—anything with a continuous spectrum of values, such as a positioning offset or a color value, is better than anything with a discrete set, such as `display` or `text-decoration`.

Now let's look at the properties we applied to `h1#pageTitle`:

```
-webkit-animation-name: throbbing;
-webkit-animation-duration: 6s;
-webkit-animation-direction: alternate;
-webkit-animation-timing-function: ease-in-out;
-webkit-animation-iteration-count: infinite;
```

`Name` and `duration` should both be fairly self-explanatory, while we've covered `timing-function` already. `Iteration-count` defaults to 1, while `direction` defaults to `normal`—by setting it to `alternate`, we make the animation go backwards on every other iteration. It spices things up a little.

As far as syntax goes, that's more or less it—but combined with the built-in transformation functions of CSS3, you can create some complex and beautiful custom animations.

Have a go hero

With a feature like CSS3 animations, the best way to learn about it is to get your hands dirty and play around with the code itself.

How would you go about applying custom animations programmatically? If you look back at the code we wrote to display our modal tweet view, can we modify that to use a custom animation? Is there an easy way to parameterize the animations that get applied?

Try to investigate defining new animations on the fly—is this possible? What are some effects you can create by doing this? You can look online for some exciting examples of the possibilties; there are a number here, for instance: `http://webdesignerwall.com/trends/47-amazing-css3-animation-demos`

Animations: CSS3 or HTML5?

In the previous chapter we learned all about the canvas element in HTML5 browsers for procedural drawing directly to the browser, and the flexibility of that API. We also noted that there is support in many mobile browsers (unfortunately, not including Android at the time of writing) for Scalable Vector Graphics, SVG. One could be forgiven for getting a little confused; with all of these graphical APIs already in the browser, why do we need these new CSS3 features?

That's an interesting question, and I'm glad that I thought to write it. In a lot of cases, there are some clear divisions between what belongs to CSS and what belongs to these JavaScript APIs. If you need to manipulate some elements on the page, such as we were doing with our modal Tweet view, then CSS transitions are clearly the best solution. Likewise, if we have complex drawings we wish to render on the fly, then either canvas or SVG is the best solution; right now, there's no CSS API for drawing, as such.

There are edge cases, however, particularly for animating simpler shapes or manipulating pre-existing images. Take our resoundingly unattractive dashboard, for instance: had we defined our red and yellow as DOM elements, rather than paths on the canvas, it would have been quite easy to animate them—perhaps sliding up as new foods are added, and fading away when foods are removed. This would be possible on the canvas as well, but more custom code would be required, either through a library or written by ourselves, to get the animation to look just right.

In cases where there is some dispute, the two important issues are:

1. How does each API work on your target devices?
2. How much do you like each API?

Right now, for performance, it's a bit of a wash: Apple devices have hardware-accelerated 3D transforms, and questionable performance with canvas animation; Android devices don't hardware accelerate anything, but tend to do quite well with the canvas; and the latest BlackBerry handsets have great SVG support, and decent performance for the other two options.

So for most developers it will come down to personal preference: do you prefer procedural definitions for animations and the like, or declarative ones? In my opinion, the CSS3 syntax gives excellent support for the majority of animations you will want to perform for a mobile application, even if the canvas offers more for flexibility and customization. But the new HTML5 APIs are well worth learning in their own right.

Summary

In this chapter, we have played around with the new CSS3 features in a few interesting ways:

- Using built-in transforms to perform a simple, modal view transition
- Integrating the iScroll mobile scrolling library to mimic fixed positioning on a mobile HTML page
- Defining our custom animations using the CSS3 explicit keyframes syntax

In the next chapter, now that we have a firm grasp of the state of the art for mobile web development, we will start to look at the custom APIs that PhoneGap provides, beginning with access to device sensors.

7
Accessing Device Sensors with PhoneGap

The preceding chapters will have given you a good sense of the type of application that PhoneGap enables you to write: stateful, native-style mobile applications that use HTML and the latest JavaScript and CSS APIs to provide rich functionality and a good user experience. With this grounding, we can now start to look at the APIs that are provided by the PhoneGap framework itself.

Beginning with this look at the PhoneGap's device sensor APIs, we will start to write code that will only work in a PhoneGap environment—that is, less of our development process will be suitable for a desktop browser such as Safari. While PhoneGap follows W3C specifications where available–for example, with its geolocation API—most PhoneGap APIs are not available in any browser, so you will need to test on a device, or at least an emulator.

In this chapter, we will look at the device sensor APIs in PhoneGap—those APIs that allow developers to investigate the physical state of the device. In particular, we will learn how to:

- Query the device's geolocation API to get a user's latitude and longitude
- Monitor the accelerometer readings to detect device motion, and detect if the user is shaking their phone
- Use CSS media queries to tailor our layout based on device orientation

What are device sensors?

A device sensor is a hardware feature that may or may not be present on any given device, which allows programmatic access to information about the device's physical surrounding.

Sensors have become increasingly important to web developers over the last few years, as mobile devices have become increasingly prevalent among users. A user with a laptop computer may change their computer's location quite regularly, for instance, but they're unlikely to tilt it while they read their email, or try to point it north. Most standard computers (until quite recently) have never had these sensors as hardware features, let alone APIs exposed to web developers.

PhoneGap, at the time of writing, supports four sensor API across platforms: geolocation, accelerometer, magnetometer (or compass), and orientation. There's no hard limit on this: more devices are beginning to ship with embedded gyroscopes, so PhoneGap will likely expose that reading as well, but the four key sensors are pretty ubiquitous on modern mobile devices.

With older devices, tread with caution: older BlackBerries, in particular, have limited support for anything other than geolocation (not surprisingly—the form factor is unsuited for the landscape orientation, for one thing).

A final note on sensors in general, before we move on to writing some code: the PhoneGap sensor APIs are quite consistent between themselves, and if you can figure one out, you'll get the others. Here's the basic pattern:

```
navigator.sensor.getCurrentVariable(successCallback, errorCallback,
options)
```

This function asynchronously queries the sensor for the current reading. Once the reading is made, it is passed as the first parameter to the function `successCallback`.

```
navigator.sensor.watchVariable(successCallback, errorCallback,
options)
```

This periodically calls `getCurrentVariable`, passing the result to `successCallback` each time. The interval is specified in the last parameter, `options`. The `watchVariable` will return a `watchId` variable, which you would use for:

```
navigator.sensor.clearWatch(watchId)
```

This ceases the watch (and periodic polling) initiated by `watchVariable`. Let's put all this into practice now, and try out the geolocation API.

Time for action – A postcard writer

We're going to start a brand new application here: it will allow the user to write messages, and then label those messages based on where they are from. Set up your environment to have a new project space, copy over `xui.js` and `mustache.js` from the **Food List** application, and create a new `index.html` file:

1. Let's start by filling our `index.html` with some default markup—a text box to write new postcards in, and a list of postcards to display:

```
<html>
    <head>
        <title>Postcards from Thumbs</title>
        <link rel="stylesheet" href="style.css" />
        <meta name="viewport" content="width=device-width,
         initial-scale=1.0" />
    </head>
    <body>
        <div id="main">
            <h1 id="pageTitle">Postcards</h1>
            <form id="newPostcard">
                <textarea id="postcardContents"></textarea><br>
                <button id="postTheCard">Post It</button><br>
            </form>

            <div id="postcardsPosted">
                <div class="postcard">
                    <blockquote id="contents">This is a sample
                     postcard, that I have written from sunny
                     Vancouver.</blockquote>
                    <aside>Vancouver, B.C.</aside>
                </div>
            </div>
        </div>
    </body>
    <script src="phonegap.js"></script>
    <script src="xui.js"></script>
    <script src="mustache.js"></script>
    <script src="app.js"></script>
</html>
```

Incidentally, note that we're using the HTML5 `aside` tag to mark up the address of the postcard. It shouldn't affect the layout—in terms of style, it's a regular block shaped element on the page—but it gives a bit of a neater semantic markup to our application.

The overall layout is fairly similar to our Food List application (hey, I never said I was a designer). One key change is that we're including `phonegap.js` in a `script` tag—as I've mentioned, this is not essential for every PhoneGap application, but we will be accessing the PhoneGap sensor APIs with this application, so we need the JavaScript file included on the page. Remember that your `phonegap.js` file will automatically be generated by PhoneGap when your application is built, and will be different for each platform you're deploying to.

2. The next step is to style things; here's a basic stylesheet that should give us something to work with:

```
body {
    background: #D9F8BA;
    font-family: Helvetica, Arial, sans-serif;
}
#newPostcard {
    width: 300px;
}
#postcardContents {
    width: 100%;
    height: 100px;
}
#postTheCard {
    margin: 20px;
    float: right;
}
#postcardsPosted {
    clear: right;
}
.postcard {
    clear: both;
    margin: 30px;
    width: 270px;
}
.postcard aside {
    font-style: italic;
    float: right;
}
```

Save that as `style.css`, and load everything in a browser:

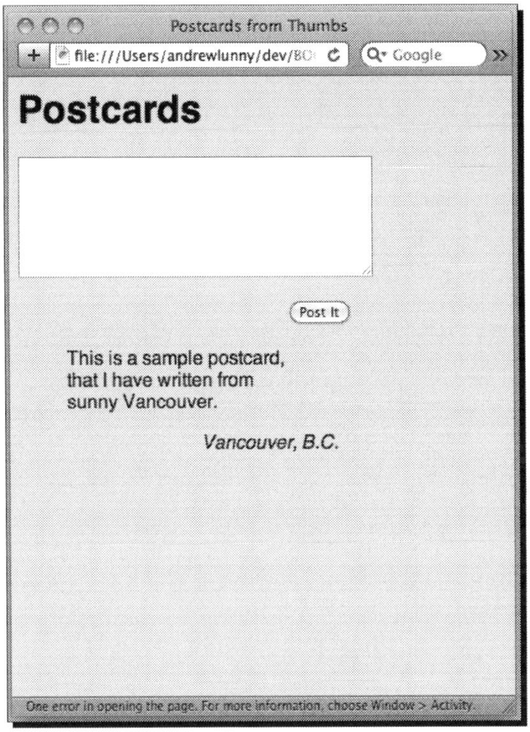

3. Now let's start to figure out the behavior. When the user submits a postcard through our form, we want to check where it came from, and display it at the top of the list. Let's start with the last part: what to do once we have the text of the postcard and the location that it came from. Open a new file called `app.js` and add the following:

```
var postcardTemplate = '<div class="postcard"><blockquote
class="contents">' +
    <{{ text }}</blockquote><aside>{{ place }}</aside></div>';

function postcardMarkup(text, place) {
    return Mustache.to_html(postcardTemplate, {
        text: text, place: place
    });
}

function appendNewPostcard(text, place) {
    x$(<#postcards-posted').top(postcardMarkup(text, place));
}
```

You'll remember **Mustache**: the templating library we had been using in the previous application. We set up a mustache template, `postcardTemplate`, that contains the markup we need. Then we have a wrapper function, `postcardMarkup` that takes a postcard and a location and spits out the templated markup. Finally, `appendNewPostcard` wraps `postcardMarkup` with **XUI** to add the new postcard to the DOM. Simple, no?

To test that everything works correctly, let's remove the original postcard content from our `index.html` page and add it dynamically, on the `DOMContentLoaded` event, in `app.js`:

```
x$(document).on('DOMContentLoaded', function (e) {
    var content = ‹This is a sample postcard, that I have written
from sunny Vancouver.',
        location = ‹Vancouver, B.C.';

    appendNewPostcard(content, location);
});
```

Reload the browser; everything should look the same.

4. Next, we want to handle the input from the text area, by appending it to our list. Let's write a function called `handleNewPostcard` to do that:

```
function handleNewPostcard() {
    var postcardBox = x$(‹textarea#postcardContents')[0],
        content = postcardBox.value;

    postcardBox.blur();
    appendNewPostcard(content, ‹Terra Incognita');
    postcardBox.value = ‹';
}
```

We blur the text area to ensure the device's hardware keyboard is hidden, and then append the contents to the list. Right now, we haven't queried the user's location, so we just list it as *Terra Incognita*.

5. Now to add the location. As mentioned above, PhoneGap follows closely the HTML5 Geolocation API, so we can get an initial test going in our desktop browser (assuming that the browser supports the HTML5 API). Let's modify `handleNewPostcard` to make the correct API call:

```
function handleNewPostcard() {
    var postcardBox = x$(‹textarea#postcardContents')[0],
        content = postcardBox.value;

    postcardBox.blur();
    navigator.geolocation.getCurrentPosition(function (resultsObj)
{
```

```
        appendNewPostcard(content,
        positionToLatLongString(resultsObj));
    }, function () {
        appendNewPostcard(content, ‹Terra Incognita');
    });
    postcardBox.value = ‹';
}

function positionToLatLongString(pos) {
    return pos.coords.latitude + ‹,' + pos.coords.longitude;
}
```

A couple of things to note:

- We've created a wrapper function, `positionToLatLongString`, to convert the position object that is returned to a human readable string
- If the geolocation call is unsuccessful, we call `appendNewPostcard` anyway with the default location in place

Load Safari, enter some text in the text area, and then call `handleNewPostcard` from the console:

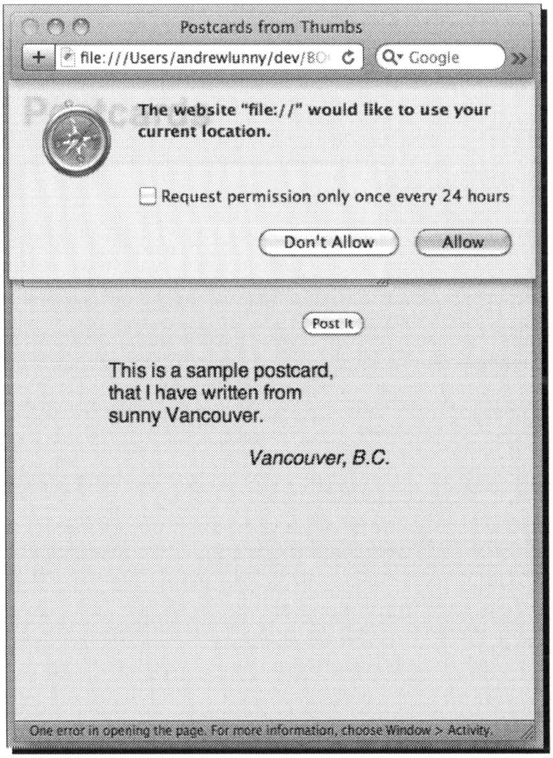

6. Whether using PhoneGap or the standard browser APIs, any application that attempts to access the Geolocation API requires manual approval from the user. If we click **Don't Allow**, the error callback is called (in our case, adding the postcard from *Terra Incognita*). Otherwise, the success callback is called.

7. So we've gotten a geolocation reading, but, I think you'll agree, it's not the most user-friendly experience. Unless I've been writing postcards incorrectly all of this time, most people don't sign off by providing their latitude and longitude coordinates.

   The most common use for location readings is to draw a map centered on the user's location. In our case, we're going to use a **reverse geocode** API to convert the location reading into something more readable. We'll use the **Google Geocoding API** for this, which you can read about here at `http://code.google.com/apis/maps/documentation/geocoding/`.

   Let's begin to do this. First, we'll write a very simple function that takes a latitude-longitude string and converts it to a Google Geocoding URL:

```
function urlForLatLong(latlng) {
    return ‹http://maps.googleapis.com/maps/api/geocode/
json?latlng=' +
        latlng + ‹&sensor=true';
}
```

Next, we need a function to query that URL and then parse the results. Here's what I came up with:

```
function getHumanLocation(latlng, callback) {
    x$().xhr(urlForLatLong(latlng), function (e) {
        var data = JSON.parse(this.responseText);

        if (data.status != "OK") {
            return callback("Terra Incognita");
        }

        var result = data.results[0],
            output,
            locality = ‹',
            region = ‹',
            country = ‹';

        result.address_components.forEach(function (part) {
            if (part.types.indexOf(‹locality') >= 0) {
               locality = part.long_name;
            } else if
              (part.types.indexOf(‹administrative_area_level_1')
               >= 0) {
                region = part.short_name;
            } else if (part.types.indexOf(‹country') >= 0) {
                country = part.long_name;
            }
        });

        output = [locality, region, country].join(‹, ‹);

        callback(output);
    });
}
```

One thing to note is that Google's API returns us a top-level value called `status`: if this value equals `OK`, then we can proceed; otherwise, we'll halt the execution right away.

We could use `result.formatted_address` to get the street address—our code goes through a few more backflips to get a nicer sign-off.

8. Finally, let's change `handleNewPostcard` to take advantage of our new functions:

```
function handleNewPostcard() {
    var postcardBox = x$(‹textarea#postcardContents')[0],
        content = postcardBox.value;
```

```
        postcardBox.blur();
        navigator.geolocation.getCurrentPosition(function (resultsObj)
    {
        var latlng = positionToLatLongString(resultsObj);

        getHumanLocation(latlng, function (placename) {
            appendNewPostcard(content, placename);
        });
        }, function () {
            appendNewPostcard(content, ‹Terra Incognita');
        });
        postcardBox.value = ‹';
    }
```

Test the round-trip once more and, ta-da! Success!

What just happened?

We took some of the techniques that were familiar from earlier chapters—in particular, templating with `Mustache.js` and using XUI for both DOM manipulation and event binding—and combined those with the use of a device sensor API to produce a more involving experience for the user. Combined with a remote geocoding API, we were able to add a personal touch to our, admittedly, plain looking application.

A latitude and longitude reading is not terribly interesting in and of itself, so most PhoneGap applications will end up passing that data somewhere else. The most likely destination is a mapping application: either rendering a map on the device, through one of the PhoneGap map plugin, or using a JavaScript API to receive a map back. Many other remote APIs—for example, Twitter's—can accept geolocation data, typically for reverse geocoding on their own servers.

PhoneGap versus HTML5

While developing in Safari, we were able to use the Geolocation API from HTML5, rather than using PhoneGap's APIs. What's the difference?

The main factor is cosmetic; using PhoneGap, we can control the prompt that is shown to the user. Since PhoneGap, as we've covered, serves your application through a web view on the device, queries for the location service from JavaScript are interpreted as being from `file://storage/long-complicated-uuid/index.html`, rather than **My Pretty App Name**.

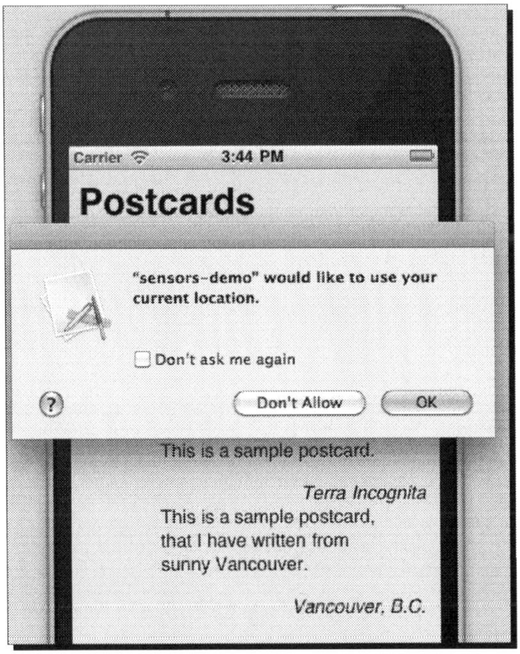

In terms of functionality, there is not much difference between PhoneGap and the standard HTML5 approach; in fact, PhoneGap has targeted compatibility with this kind of API as much as possible. But whereas the behavior of the user-facing prompt can be unpredictable depending on the device in use, PhoneGap standardizes what the user sees, across platforms.

Other geolocation data

In the example above, we only used the latitude and longitude properties of the position object that we received. There are some other attributes that can be accessed:

- Altitude: the height of the device, in meters, above sea level
- Accuracy: accuracy level of the latitude and longitude, in meters
- AltitudeAccuracy: you can probably figure this one out
- Heading: direction of travel, in degrees clockwise from true north
- Speed: current ground speed of the device, in meters per second

That said, latitude and longitude are far and away the best supported of these properties, and the ones that will be most useful when communicated with remote APIs. The other properties are mainly useful if you're developing an application for which geolocation is a core component of its standard functionality.

The accuracy property is the most important of these additional features. Location coordinates can be measured from the WiFi network the device is connected to, the cellular towers the device is accessing, or a GPS receiver on the device itself. As an application developer, you typically won't know which particular sensor is giving you the location data, so it may be important to correct for low accuracy readings.

Pop quiz: Geolocation

1. When is the user prompted to allow geolocation access?
    a. When the application is initially loaded
    b. When the application calls `getCurrentPosition`
    c. When the location data is passed to a remote API
2. What happens when the user refuses to allow access to their location data?
    a. The error callback is called
    b. The callback is called, with the error as the first parameter
    c. The callback is not called, and the application continues

3. How is geolocation data calculated?
    a. From the user's network connection: either through WiFi or the cellular network
    b. By the GPS sensor in the device
    c. All of the above

Accelerometer data

A device's accelerometer is a simple sensor that is almost ubiquitous among modern mobile computing devices, that measures the position of the device in 3D space. It is possible to get the X, Y, and Z coordinates for the position of a device, and then to track those coordinates through time. If you've played a mobile racing game, or any game involving balancing or leaning, you must have used your device's accelerometer.

While geolocation is not only in HTML5, but quite prevalent on desktop browsers these days, access to the device accelerometer is almost exclusively a mobile concern, at present. Laptop or desktop computers stay in one place during use, for the most part, while mobile devices are often moved from side to side, perhaps even unwittingly.

Browser vendors are beginning to introduce accelerometer access as JavaScript APIs—in particular, Apple's iOS devices introduced a `DeviceMotionEvent` in version 4.2 that fires every time the device has been moved. However, for cross-platform development, PhoneGap's `navigator.accelerometer.getCurrentAcceleration` is currently the most predictable and reliable way to track how the device is moving.

Time for action – Detecting shakes

Unlike the geolocation API, a single reading of the device's accelerometer is not all that interesting. In this example, we're going to poll the accelerometer for a period, to see how the readings change as we shake the device.

You will not get very far with the accelerometer on a desktop web browser or an emulator. This is the first API we've encountered that really requires a device at hand to measure your progress. You can refer back to *Chapter 1* for details on how to get a PhoneGap application loaded onto your device.

1. Let's go back to our Postcard application, and set up a watch on the accelerometer. To do this, we'll have to listen for the `deviceready` event and, once it fires, call `watchAccelerometer`:

```
x$(document).on('deviceready', function () {
    navigator.accelerometer.watchAcceleration(function (reading) {
        if (reading.is_updating) return;
```

```
            console.log("x: " + reading.x + "; y: " + reading.y +
                "; z:" + reading.z);
    }, null, { frequency: 300 });
});
```

If the reading is updating when we poll it, we return from the function and wait for the next run. Otherwise, we log the current measurements to the console. The function is called every 300 milliseconds.

Here's a sample set of readings I saw when running on my device (the output is from Android's `logcat` logger, hence the format), holding it quite still:

```
D/PhoneGapLog(32642): file:///android_asset/www/app.js: Line 121 :
x: -1.334794044494629; y: 4.399372100830078; z:8.662541389465332
D/PhoneGapLog(32642): file:///android_asset/www/app.js: Line 121 :
x: -0.46309182047843933; y: 4.521955490112305; z:8.580819129943848
D/PhoneGapLog(32642): file:///android_asset/www/app.js: Line 121 :
x: -0.5039528608322144; y: 4.562816619873047; z:8.68978214263916
D/PhoneGapLog(32642): file:///android_asset/www/app.js: Line 121 :
x: -0.46309182047843933; y: 4.521955490112305; z:8.853225708007812
D/PhoneGapLog(32642): file:///android_asset/www/app.js: Line 121 :
x: -0.6537767052650452; y: 4.399372100830078; z:8.771504402160645
D/PhoneGapLog(32642): file:///android_asset/www/app.js: Line 121 :
x: -0.5311935544013977; y: 4.399372100830078; z:8.730643272399902
D/PhoneGapLog(32642): file:///android_asset/www/app.js: Line 121 :
x: -0.46309182047843933; y: 4.44023323059082; z:8.771504402160645
```

The particular numbers are not too important for our purposes—we're more interested in how they change between readings.

Here's another set of readings, while I'm shaking the device:

```
D/PhoneGapLog(32642): file:///android_asset/www/app.js: Line 121 :
x: -2.3426997661590576; y: 4.903325080871582; z:8.662541389465332
D/PhoneGapLog(32642): file:///android_asset/www/app.js: Line 121 :
x: -0.46309182047843933; y: 6.047434329986572; z:8.281171798706055
D/PhoneGapLog(32642): file:///android_asset/www/app.js: Line 121 :
x: 6.973618030548096; y: 11.304888725280762; z:7.818079471588135
D/PhoneGapLog(32642): file:///android_asset/www/app.js:
Line 121 : x: -19.504337310791016; y: -13.375181198120117;
z:3.296124219894409
D/PhoneGapLog(32642): file:///android_asset/www/app.js: Line 121 :
x: 18.387470245361328; y: 4.399372100830078; z:-1.5663399696350098
D/PhoneGapLog(32642): file:///android_asset/www/app.js:
Line 121 : x: -18.046960830688477; y: -11.876943588256836;
z:13.715690612792969
D/PhoneGapLog(32642): file:///android_asset/www/app.js: Line 121 :
x: 1.375655174255371; y: 10.800935745239258; z:19.150209426879883
```

2. We can see the difference in acceleration is quite high: as much as 30 degrees difference on a single axis, in some cases.

3. To detect shakes, we're going to decide on a threshold, and check whether all of axes change by that threshold between readings. Here's the code to do this:

```
x$(document).on('deviceready', function () {
    var previousReading = {
        x: null,
        y: null,
        z: null
    }

    navigator.accelerometer.watchAcceleration(function (reading) {
        var changes = {},
            bound = 3;

        if (previousReading.x !== null) {
            changes.x = Math.abs(previousReading.x, reading.x);
            changes.y = Math.abs(previousReading.y, reading.y);
            changes.z = Math.abs(previousReading.z, reading.z);
        }

        if (changes.x > bound && changes.y > bound && changes.z >
        bound) {
            console.log(‹shake detected');
        }

        previousReading = {
            x: reading.x,
            y: reading.y,
            z: reading.z
        }

  }, null, { frequency: 300 });
});
```

There are a few things to note here. Firstly, we're storing the previous reading from the accelerated as a local variable, stored in `deviceready` handler's function scope. We then define a bound for each dimension (in this case, I'm setting it to three degrees), and check that every dimension changes by that bound. If so, we log **shake detected**—and once we've detected it, we can hook in more complex shake handling.

These bounds are of course, just one estimate—your own testing may suggest higher bounds, or different conditions per dimension.

4. Now let's write the `shakeHandler` function. We want the shake handler to be, essentially, an *undo* button—it will prompt the user to delete their last postcard, and then remove it from the DOM if they select it. Let's write that function now:

```
var canPrompt = true;

function shakeHandler() {
    var msg = ‹Would you like to delete your last postcard?';

    if (canPrompt && confirm(msg)) {
        x$(‹div.postcard:first-child').remove();
    }

    canPrompt = false;

    window.setTimeout(function () {
        canPrompt = true;
    }, 5000);
}
```

We set up a global variable called `canPrompt` that checks whether or not we should prompt the user—we don't want to show the confirm dialogue more than every five seconds. The shake event we've set up may fire multiple times from a single shake by the user, so five seconds is a good limit for the user's shake.

After that, it's fairly simple XUI—use a CSS3 selector to grab the topmost postcard, and remove it from the DOM.

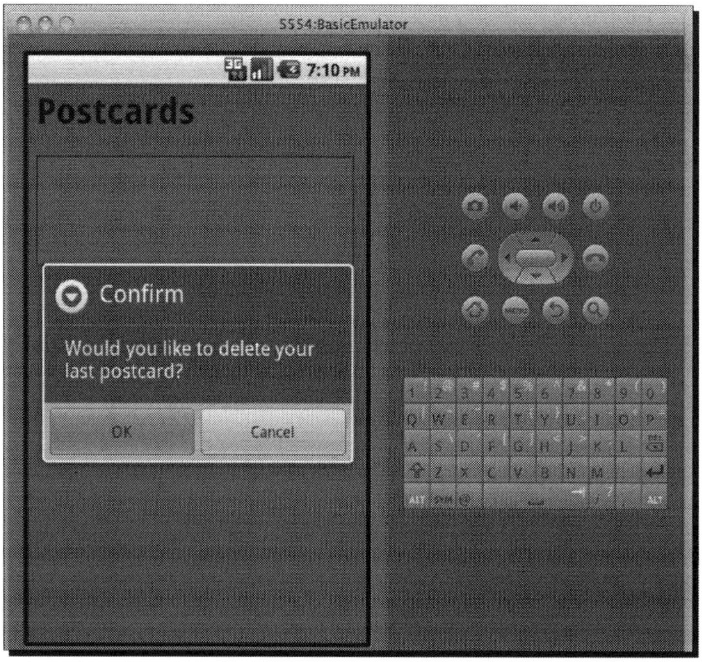

What just happened?

We were able to use PhoneGap's device accelerometer API to detect whether the device is shaking or not, and then alter behavior based on what we detected.

This is a fairly common use of the accelerometer API: the device axes are fairly low-level data to work with, so you will need to write some code that translates these readings into a higher level abstraction, whether that's a shake or a tilt or whatnot. It is likely that as these APIs come to the fore in the browser, there will be higher-level libraries that abstract away the detail of accessing these readings directly—until then, we can enjoy playing with the raw data.

Device orientation and device motion events

There are two related specifications for browser-based acceleration data: the **DeviceOrientationEvent** and the **DeviceMotionEvent**. I'll quote from the specification itself, which can be found at `http://dev.w3.org/geo/api/spec-source-orientation.html`:

> *This spec provides two new DOM events. The first [DeviceOrientationEvent] is a simple, high-level source of information about the physical orientation of a device. While the spec is agnostic to the source of information, this is typically implemented by combining information from an accelerometer and a magnetometer. The second event [DeviceMotionEvent] provides direct access to motion data from an accelerometer and gyroscope and is intended for more sophisticated applications.*

At the time of writing, the DeviceOrientationEvent is supported on a few desktop browsers such as Google Chrome (assuming the computer has the appropriate sensors to expose this data), while the DeviceMotionEvent fires on iOS 4.2 and above devices. It is likely that all major mobile browsers will expose both APIs in the next couple of years.

The PhoneGap acceleration API is closer to the DeviceMotionEvent, but does not line up exactly with the specification at the moment. It's unclear at the moment how this API will change: the motion events fire very often in browsers that implement them, and firing these continually in every PhoneGap application would have a considerable performance impact.

Have a go hero

Now that we've interpreted some data from the accelerometer, can you detect other high-level information about the device from the readings? What should the readings look like when the device is tilted to either side, or back and forwards?

A good way to play around with this is to combine the accelerometer readings with the CSS3 `rotate` transform: can you get an element on the screen to react based on how the device is held?

Orientation media queries

The passage we quoted from the DeviceMotion specification highlighted that there are simple and complex uses of device sensor data: for complex applications, you will want low-level readings of how the device is held in 3D space, while for most applications you will just want to do a little design tweak.

Thankfully, mobile browsers include support for CSS3 media queries on the device that allow for powerful contextual design based on the properties of the device. iOS browsers above version 4.0 include `orientation` queries, so we can directly specify how an application should look in portrait mode and in landscape mode, while almost all modern mobile browsers support `min-width` and `max-width` queries. If we're able to make a few assumptions about the device itself, we can quite easily use these queries to get the same orientation-specific styling.

Time for action – Landscape postcards

As mentioned above, there are a few slight changes between the media queries we'll need for different platforms; we're going to get things working for iPhone, and then make the necessary changes to have the same effect on Android:

1. First things first, let's see where we are at the start. Set up your application as an iPhone project, and launch it on the iPhone Simulator. Hit *Command* and *Left* to orient the device in landscape mode:

It looks... like our existing interface, turned to the left. Not ideal.

2. We need to enable rotation of the application: by default, iOS PhoneGap applications have this turned off. Edit your application's `plist` file to ensure that **Supported Interface Orientations** includes all the four options:

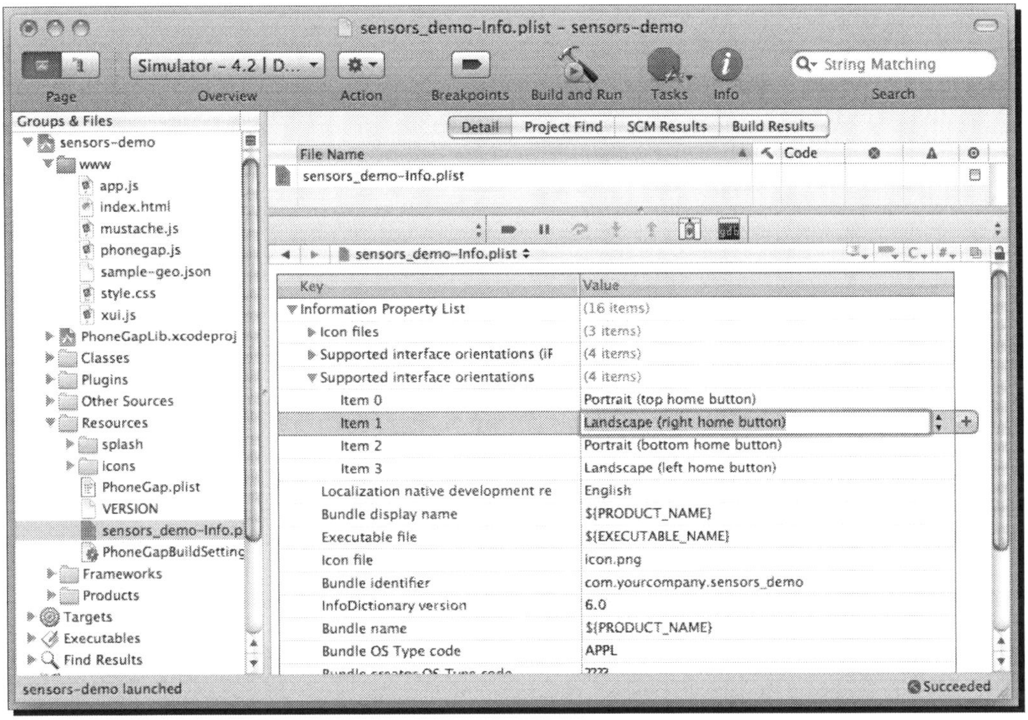

Now let's launch the simulator again and check how it looks:

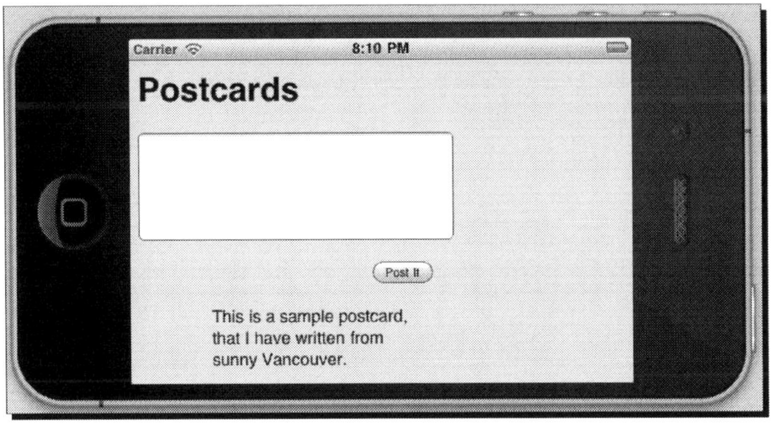

That's closer to what we're after, but it would be nice to have the list of postcards aligned to the right of the `textarea`. Let's do that.

3. Go into `style.css` and add the following:

```
@media all and (orientation:landscape) {
    #newPostcard {
        width: 200px;
        float: left;
    }
    #postcards-posted {
        clear: none;
        float: left;
    }
    .postcard {
        clear: none;
        margin: 20px;
        width: 220px;
    }
}
```

What we're doing here is querying the orientation of the device, to see whether it's in the landscape mode. If it is, we override some of the styles that we had defined above. Simple, no? Fire up the iPhone Simulator again and have another look:

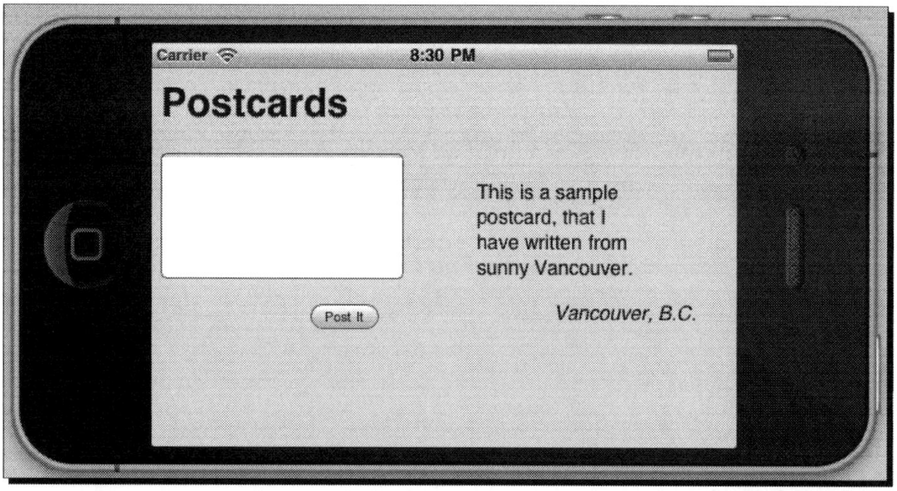

Fantastic! The next step you might want to do is pull in iScroll, and get the list of postcards scrolling on the right-hand side. But let's not worry about that for now.

4. Now let's throw our code into an Android project and give it a shot. To rotate the Android emulator, press *CTRL-F11* (note that this is an Android emulator running Android 2.2—older releases may not support these media queries):

So much empty space!

5. As you've probably guessed, orientation queries aren't supported on Android. Luckily, we can query the `device-min-width` for a similar effect; if the width is bigger than, say, 400 pixel, we're almost certainly in landscape mode:

```
@media (orientation:landscape), (min-device-width: 400px) {
    #newPostcard {
...
}
```

Note that we're using the standard CSS comma not the word "or".

Also note that this approach is more fallible than the orientation query: this particular query would fail on a tablet, for instance. But for now, it serves our purpose. Reinstall the application and fire it up once more:

Success! A fluid, responsive, cross-platform design.

What just happened?

We were able to use CSS3 media queries to customize the design of our application in a simple, declarative fashion, without having to subscribe to any JavaScript events or access any PhoneGap orientation APIs.

It may seem like a bit of an anti-climax—oh, we just write more CSS?—but that's what's especially nice about media queries: their simplicity. Our presentation code is kept cleanly away from our behavior, with everything in its right place. Since styling content depending on orientation is such an essential component of mobile design, it's great that this ubiquitous use case can be abstracted away with so little effort.

Other media queries

There are a whole host of other things that can be selected for, including the device's resolution, its DPI, which is especially useful for high resolution displays, such as the iPhone 4, it's aspect ratio, even the color output quality of the display. While not all of these options are implemented everywhere, and some aren't implemented anywhere, it's very encouraging for the future of mobile development and design that so many options are available.

Pop quiz: Orientation and media queries

1. What is the main benefit of CSS media queries?
   a. Reduce the size of files: clients only download the styles that apply to their device
   b. Target different experiences with a high level of granularity
   c. Write more CSS and less JavaScript
2. Why is `orientation` preferred to `min-device-width`, in our example?
   a. `orientation` can change during a user's session, while `min-device-width` is fixed
   b. `orientation` is a more precise measure, while `min-device-width` can be unpredictable on different devices
   c. `min-device-width` depends on an inference, while `orientation` tells us what we want to know
3. Is it necessary to support all orientations on a device?
   a. No: you can configure each platform to allow or deny certain orientations
   b. Yes: if you do not explicitly support different orientations, a weak substitute is provided instead

Magnetometer: The missing API

The adroit readers of this book will have noticed that there's another PhoneGap device sensor we have not covered: the magnetometer, or, to give it's better known name, the humble compass.

Truthfully, it's not terribly interesting. The PhoneGap magnetometer API is a simple one almost identical to the accelerometer, in fact. If you can follow the accelerometer example above, you can certainly extend your applications to include the compass. And generally, it's not all that useful on its own; unless you're writing a competitor to Google Maps, it's questionable how much you will need the compass access. But trust me: the API is there, it works well, and it will show you the right way home.

Summary

In this chapter, we have looked at the device sensor APIs available to PhoneGap developers, including:

- Location data, with `navigator.geolocation.getCurrentPosition`
- The three-dimensional position of the device, through the accelerometer data
- Orientation of the device, using CSS queries

Now that we've whetted your appetite with a taste of some of the native capabilities available to PhoneGap developers, the next chapter will look at the ever-popular camera API.

8

Accessing Camera Data and Files

Mobile devices, almost ubiquitously, have one or two cameras mounted on them, and the photographs associated with them are some of the most obvious user data for applications to access. While camera access in mobile browsers is still nascent, PhoneGap has long had support for accessing camera data and files through applications. Although there's just a single JavaScript that gets used—`navigator.camera.getPicture`*—the myriad use cases for the camera API make it one of the more popular PhoneGap APIs.*

In this chapter, we will get to grips with PhoneGap's camera API and how it can be used; in particular, we'll:

- Master the `getPicture` function, and the various options that can be passed to it
- Render files from the camera directly onto our page, to be integrated with the rest of our application
- Grab the camera data directly, and save it to our local database

Let's get started right away, by seeing what we can do with the `getPicture` function.

Time for action – Hello World with the Camera API

As you may well be aware, there is no way to use anything like the camera API from JavaScript in modern browsers (for now, at least). All the examples in this chapter will need to run on a device, although you can get started on an emulator.

We're going to start with the postcard application we developed in *Chapter 7*, and add functionality for the user to access her camera; it will become a Picture Postcard application.

1. Firstly, we'll set up the interaction for the user to choose to add a picture to their postcard. When the user holds her finger down on the text of the postcard, she will be prompted to choose a picture to add to the postcard.

2. Firstly, let's edit the mustache template in `app.js`, so we can more easily check whether the target element is the content of a postcard:

   ```
   var postcardTemplate = '<div class="postcard"><blockquote
   class="contents">' +
       '{{ text }}</blockquote><aside>{{ place }}</aside></div>';
   ```

   We've added the class attribute that will help XUI confirm that the element being touched is the one we're looking for.

3. If we're going to be handling long touches, we also need to ensure that the user's action isn't interpreted as a selection action; we don't want the user to copy the text. Thankfully, there are a couple of CSS rules we can add to ensure that these interactions are disabled, which should have no effect on desktop browsers; add the webkit specific rules to `style.css`:

   ```
   body {
       -webkit-user-select: none;
       -webkit-touch-callout: none;
       background: #D9F8BA;
       font-family: Helvetica, Arial, sans-serif;
   }
   ```

4. The next thing we need to write is the code to handle the long touches. We're going to watch for a `touchstart` event, and log the time that it was fired, along with the element it was fired on. Then, when a `touchend` event fires, we're going to check if the same element is the target, and what the duration of the touch was. If it was above half a second, we can treat the event as a long touch.

   Because the iPhone, in particular, is tricky about event delegation, we'll need to find this individually each time we add a new postcard. There are a few things going on, so make sure you keep up.

   Here's the new code we're going to add:

   ```
   var longTouch = {
       element: null,
       since: null
   }
   ```

```
function longTouchHandler() {
    alert('long touch!');
}

var touchHandlers = {
    start: function (e) {
        var now = +(new Date());
        longTouch.element = e.target;
        longTouch.since = now;
    },
    end: function (e) {
        var now = +(new Date()), duration;
        if (longTouch.element == e.target) {
            duration = now - longTouch.since;
            if (duration > 500) longTouchHandler(e.target);
            longTouch = {};
        }
    }
}

function bindLongTouchListener() {
    x$('div.postcard:first-child')
        .on('touchstart', touchHandlers.start)
        .on('touchend',   touchHandlers.end);
}
```

5. Right now, of course, we're just stubbing out `longTouchHandler`, rather than doing anything useful. Now we just need to modify our `appendNewPostcard` to call `bindLongTouchListener`, so it's bound for every postcard we append:

```
function appendNewPostcard(text, place) {
    x$('#postcards-posted').top(postcardMarkup(text, place));
    bindLongTouchListener();
}
```

6. Let's fire up the code on a device or an emulator, to see what happens.

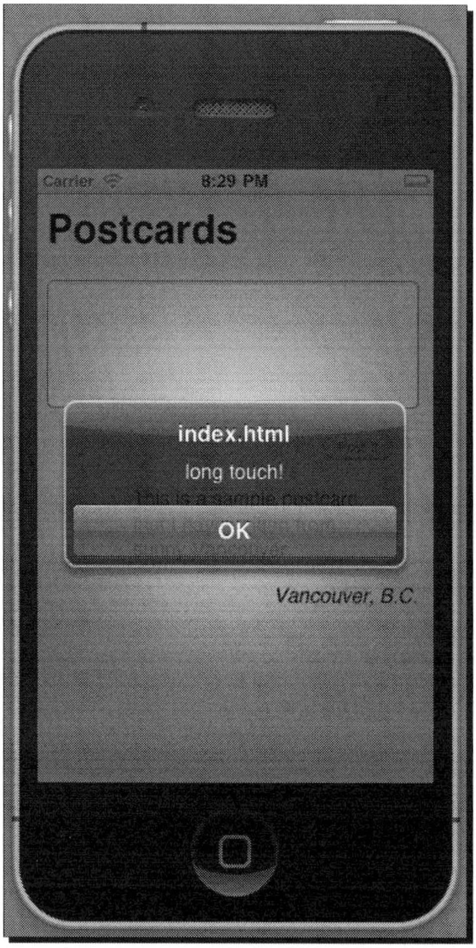

7. So there's the event handling; now for the important stuff. Let's make our first call to `getPicture`, with a couple of dummy handler functions:

```
function longTouchHandler() {
    navigator.camera.getPicture(function () {
        alert('camera success!');
    }, function (e) {
        alert('camera failure!');
    });
}
```

Now for the exciting part: load that up and see what happens.

Erm, nothing's happening on the iPhone Simulator. Ahem. Let's try looking at the logs:

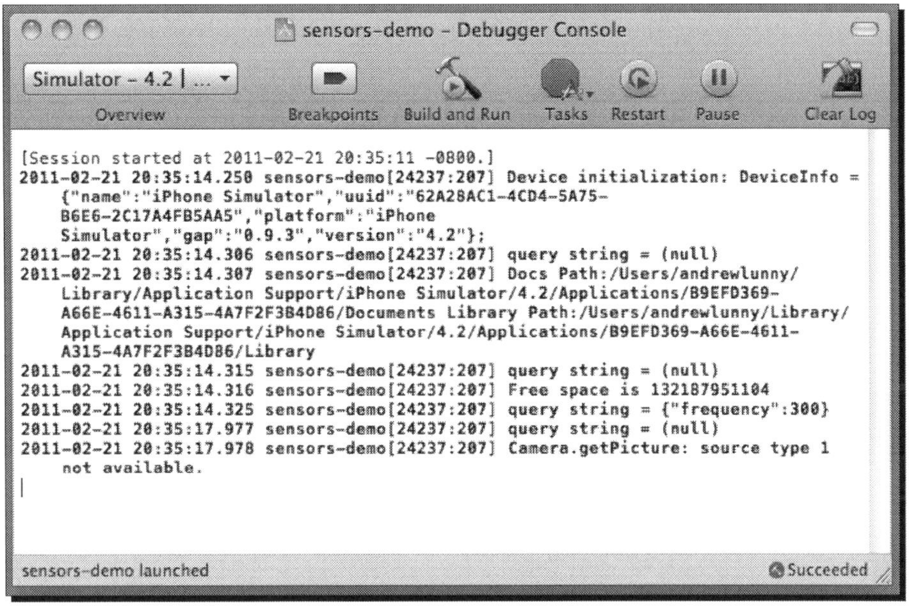

The error we're getting is: **Camera.getPicture: source type 1 not available**.

8. We'll cover the various source types later in the chapter; for now, know that the default is type 1, the Camera. The iPhone Simulator has no camera, so no dice on that count (we will cover using Android later in the chapter, but the same code should work correctly).

9. Let's modify the code accordingly. Luckily, PhoneGap defines a number of constants containing the values for the various camera source types. We're going to look in the Photo Library, as follows:

```
function longTouchHandler() {
    navigator.camera.getPicture(function () {
        alert('camera success!');
    }, function (e) {
        alert('camera failure!');
    }, {
        sourceType:
        navigator.camera.PictureSourceType.PHOTOLIBRARY
    });
}
```

That should go much better; let's try this one:

10. Well that's simulator-based development for you—no pictures available, although our Hello World purposes are basically served. Let's try the same code on my Nexus One Android phone, where I have some poorly-taken photos of my parents' cat, my girlfriend's cat, and my old roommate's eviction notice:

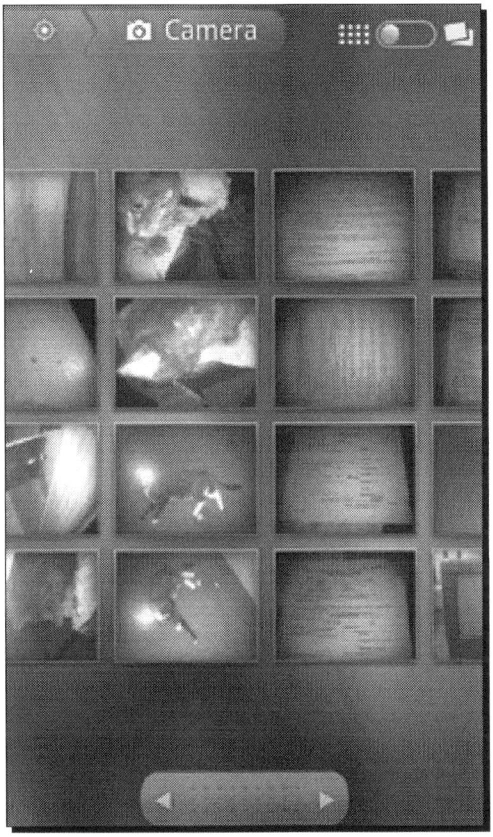

11. This isn't quite perfect—as soon as I picked an image, the application crashed with a memory warning—in most cases we need to reduce the resolution of the image before we can bring it into our application. But for a *Hello World*, we're at least getting somewhere.

    And for the sake of completeness, here's the Android version running with the camera, rather than the Photo Library as the source (with the appropriately meta-subject of this book being written):

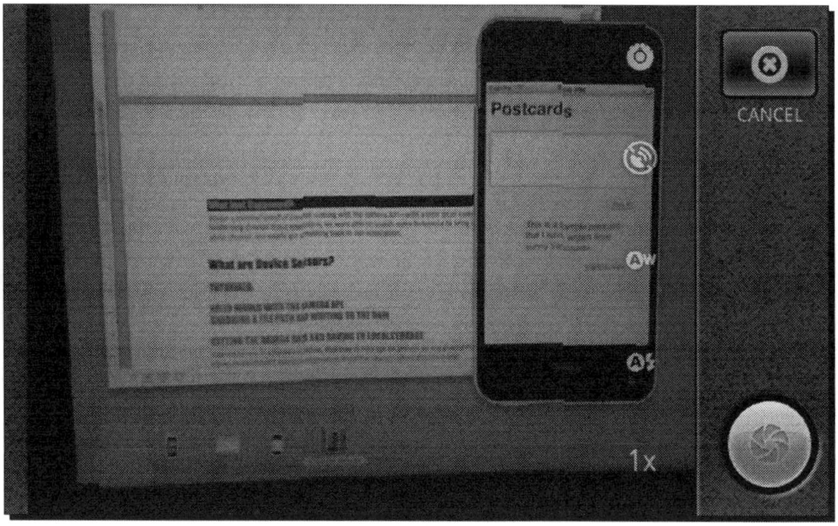

Easy enough, no?

What just happened?

We got a minimal proof of concept running with the camera API. With a little bit of code to handle long presses in our application, we were able to launch some behavior to bring up the photo chooser and nearly get something back to our application.

At any rate, the clumsy tutorial we just worked through should have been instructive in a few key ways; we've touched on these before, but this has been the most emphatic example yet.

Browsers are not emulators or devices

We've seen before that certain APIs cannot be reliably tested in a web browser, for example, when we were testing the orientation media queries, there was no way to view the difference between portrait and landscape orientations. The camera API is arguably the first one we've seen that is not very useful to test on an emulator either.

One thing we were able to tell quite easily was how PhoneGap responds when the source is unavailable—in this case, the error callback was not called, although the lack of availability was logged to the console. This will be important when we develop applications that rely on the camera functionality.

Image sources

The three major image sources PhoneGap has good support for are `PHOTOLIBRARY`, `CAMERA`, and `SAVEDPHOTOALBUM`. In case the names are not obvious, here is what those three different sources access:

- `PHOTOLIBRARY`: Images that the user has saved to her device
- `CAMERA`: An image that the user will take with her device
- `SAVEDPHOTOALBUM`: Images that the user has taken with her device

The distinction between `PHOTOLIBRARY` and `SAVEDPHOTOALBUM` is less important—both contain images that were not taken within the context of your application—than the distinction between those two and `CAMERA`. You should use `CAMERA` as the image source when your app requires, or desires, the user to take a photograph within the context of the application user experience. The other two are useful when you just want a general image from the user, for example, without the Postcards application, we don't necessarily need an image to be taken afresh.

The other major thing to be aware of is the capabilities of the devices—an older iPod Touch or first-generation iPad will not have a camera, for example, so `PHOTOLIBRARY` will be the only valid source that you can use.

Other options

Aside from `sourceType`, the most important option to pass to `getPicture` is `destinationType`, whether we want to receive a file path or the image data itself. We will cover both options later in this chapter.

`quality` is a numerical option that can be passed, with the encoded quality of the image specified (between 0 and 100). This is important when dealing with raw image data—on devices with high-resolution cameras, passing a full-quality image through the JavaScript callback could have a significant performance impact. A setting of around 50 is usually safe, though your mileage may vary.

Finally, `allowEdit` is a simple Boolean value. It is false by default, when true it gives the user a quick simple editing screen before returning the image back to the web view. At the time of writing, this option is only available on iOS devices, though it may well come on to further platforms.

Pop quiz: navigator.camera.getPicture

1. Why can we not mock `navigator.camera.getPicture` on a desktop browser?
    a. The browser camera API has a different method signature
    b. Not enough of the PhoneGap options are well enough supported
    c. There's no JavaScript access to cameras on current desktop browsers
2. Which is the most robust data source for retrieving images?
    a. `PHOTOLIBRARY`, since it's available on devices with or without an attached camera
    b. `CAMERA`, as it's supported on all PhoneGap platforms
    c. All of them: the function will fall back on whatever is available
3. How is the user interface for the image picker defined?
    a. By `HTML` and `CSS` in your `www` directory
    b. It depends on the platform's operating system
    c. It depends on the configuration of the particular device where the code is executed

What about when we finally get an image?

Once we get an image, whether from the library or directly from the camera, the most common uses are to draw the image to the document directly, or to save the image data, either locally or to a remote server. We'll look at both of these use cases, and illustrate the two different `destinationType` values that we can set.

Time for action – Getting a file path to display

1. In this example, we're going to grab a file path and display the associated on the DOM of the page. Let's edit our `longTouchHandler` from the previous chapter to do this:

```
function longTouchHandler() {
    navigator.camera.getPicture(function (img) {
        alert('camera success!');
    }, function (e) {
        alert('camera failure!');
    }, {
        quality: 50,
        sourceType:
```

```
            navigator.camera.PictureSourceType.PHOTOLIBRARY,
            destinationType: navigator.camera.DestinationType.FILE_URI
        });
    }
```

Let's grab an image and test it out:

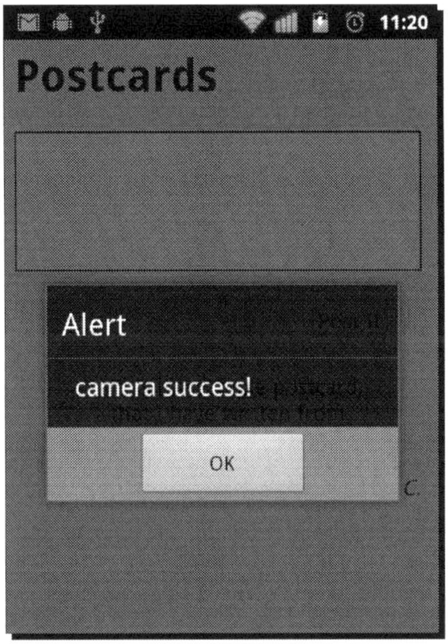

Great, the app didn't crash this time!

If you do not have a device at hand, you can try loading an image onto an SD card on an Android emulator; there are many tutorials for doing this online, such as this one: `http://www.streamhead.com/android-tutorial-sd-card/`.

2. Let's change the success handler to see exactly what we have passed:

```
function longTouchHandler() {
    navigator.camera.getPicture(function (img) {
        alert(img);
    }, function (e) {
        alert('camera failure!');
    }, {
        quality: 50,
        sourceType:
        navigator.camera.PictureSourceType.PHOTOLIBRARY,
        destinationType: navigator.camera.DestinationType.FILE_URI
    });
}
```

Build that, try again, and see what we get:

As we expected, a path—not one we would have easily guessed. Note that this path is specific to my device and my OS—you will see something different on your own device, and something very different if it's not an Android phone.

3. So let's do the obvious thing and put this image onto the DOM. First things first, we'll modify the `longTouchHandler` to draw the image out:

```
function longTouchHandler(ele) {
    navigator.camera.getPicture(function (img) {
        try {
            x$(ele).top('<img src="' + img + '">');
        } catch(e) {
            alert(e);
        }
    }, function (e) {
        alert('camera failure!');
    }, {
        quality: 50,
        sourceType:
        navigator.camera.PictureSourceType.PHOTOLIBRARY,
        destinationType: navigator.camera.DestinationType.FILE_URI
    });
}
```

Note that we've named the `ele` parameter that gets passed to the function. We've been passing it all of this time, but ignoring it up until now.

Finally, we need to add a couple of short CSS rules to style the image that gets drawn to the screen. This one should work:

```
.postcard img {
    width: 300px;
    height: 200px;
}
```

Build the app, reinstall it, and see how things look:

Looks delicious! In fact, I may prepare myself another right now.

What just happened?

Well, we managed to do something useful with the Camera API after all. By specifying the `destinationType` option to be a file path, we were able to get a value returned from the `getPicture` call without crashing the application, and we drew a picture to the screen to boot.

For simple use cases like this, the `FILE_URI` option is clearly the superior one. All we want is to draw an image to the screen, so we just need a path to that image. And since all that is being passed to our callback is a simple string, we don't need to worry about any misbehaving memory when we receive the image. Everything is tickety-boo.

Where is this image, anyway?

In the tutorial, we made a point of alerting the path of the image: `content://media/external/images/media/263`. Some things to note about that path:

- It's Android-specific: the `content://media/external` bit means the image is on the device's external storage (the SD card). Obviously this would not apply to an iPhone, for instance, where no such storage is available. And if the user switches their SD card, never mind deletes the image, you may be in trouble.
- If you do test this code on an iPhone, you may receive a path starting with `/tmp` when you alert the path. tmp means temporary! This file will have been created by PhoneGap, and may only be around as long as your application is active. If you want to persist that image data, you will have to do a bit more.

What does it all mean? Well, taking a `FILE_URI` from the camera API is not the best option if you need the image to be persisted for any significant length of time. In fact, your application has, essentially, no control over what happens to that image.

This doesn't affect silly cases like our tutorial, or a wide swath of applications that will use an image for some temporary purpose. If you do need to persist the image itself, however, there is another option you can use.

Have a go hero!

We saw a quick proof of concept of the `FILE_URI` method, but you should really put it through the ringer. What kind of file path is given on different operating systems? What happens if you save the file path to `localStorage`, and write out the image tag again when the application restarts?

Try storing an image path in this way and then deleting the image from your photo library—does anything interesting happen? Can you access the same paths from multiple PhoneGap applications—call the camera API, post the path to a remote server, and pull it down in a different application? This will vary from platform to platform—it's a fun experiment to try.

Raw image data

You want something more persistent than the `FILE_URI`, you say? You want to be able to store your images indefinitely, and send them off to remote servers? Well, my friend, you will want some raw image data.

PhoneGap gives us the option of encoding our images as base 64 data and sending those to the browser as strings. This gives us more flexibility, since we have the image itself, rather than just a pointer to it. Using the `data:` prefix for an image, source, we can do exactly what we just did with a file path, plus much more. However, it is also risky in terms of performance: when devices with 10 gigapixel cameras start to be released, our JavaScript functions will have some pretty huge strings to deal with. Let's start to dip our toes.

Time for action – Saving pictures

Don't believe that we can do the same thing as above using raw image data? Okay, I'll prove it:

1. Let's modify the `longTouchHandler` in a couple of key ways:

```
function longTouchHandler(ele) {
    navigator.camera.getPicture(function (img) {
        try {
              alert(img);
         x$(ele).top('<img src="data:image/png;base64,' + img +
         '">');
        } catch(e) {
            alert(e);
        }
    }, function (e) {
        alert('camera failure!');
    }, {
        quality: 50,
        sourceType: navigator.camera.PictureSourceType.
PHOTOLIBRARY,
     destinationType: navigator.camera.DestinationType.DATA_URL
    });
}
```

There at the bottom, we've switched our `destinationType`—as you will note, we now want a `DATA_URL`.

The other key point is the element we inserted looks a little different: instead of setting the `src` attribute of the `img` tag to whatever the success handler is passed, we prefix that data with `data:image/png;base64,`. Makes sense, right? We've also thrown in an `alert` call, to ensure we can see what we're getting passed.

Save it, build it, run it:

Yes, that looks like a base 64 encoded image alright. Hit **OK** to draw it out.

Another delicious beverage! Well I'm a regular Ernest Hemingway.

2. Yes yes, very cool, but we could do that already. Let's get a bit more creative, and save the image, so we never lose that memory.

   We're going to make a few changes to handle this. First, we'll add an `img` tag to the document to begin with, so we know where to put the saved one—add it into this spot in your `index.html` file:

```
<div id="main">
            <h1 id="pageTitle">Postcards</h1>
        <img id="postcardImage">
            <form id="newPostcard">
```

```
and modify the CSS accordingly:
#postcardImage {
    background: #CCF;
    border: 1px solid black;
    margin: auto;
    width: 300px;
    height: 200px;
}
```

   This gives us a canvas (not a `<canvas>`, mind) to draw on:

3. Now, the functions to save and load the image:

```
function saveImage(data) {
    window.localStorage.setItem('andrews.drink', data);
    return true;
}

function loadImage() {
    return window.localStorage.getItem('andrews.drink');
}
```

4. Straightforward enough, I'm sure. Now modify our `getPicture` call based on these changes:

```
function longTouchHandler(ele) {
    navigator.camera.getPicture(function (img) {
        if (saveImage(img)) {
            x$('#postcardImage').attr('src','data:image/
png;base64,' + img);
        } else {
            alert('Oh no, something spilled!');
        }
    }, function (e) {
        alert('camera failure!');
    }, {
        quality: 50,
        sourceType:
        navigator.camera.PictureSourceType.PHOTOLIBRARY,
        destinationType: navigator.camera.DestinationType.DATA_URL
    });
}
```

5. And then add some code in the `deviceready` function to populate our image if there's one already saved:

```
var img;
if (img = loadImage()) {
    x$('#postcardImage').attr('src','data:image/png;base64,' +
    img);
}
```

6. Now let's test this out; choose an image as usual:

7. Now the scary bit—delete the image, and reopen the application.

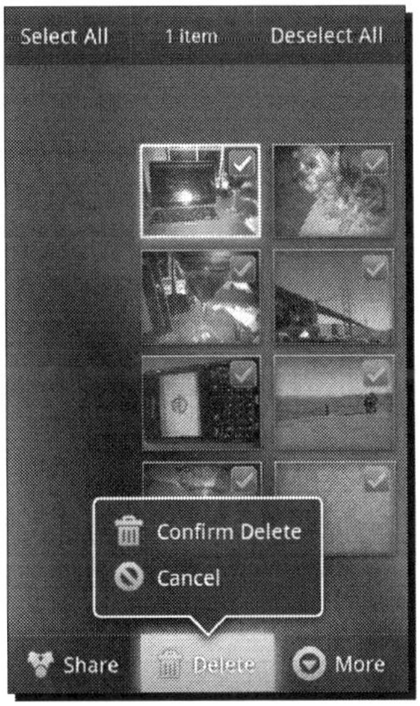

Reopen the application and it's still there! Amazing!

What just happened?

We saw the primary benefit of using the `DATA_URL` destination type as opposed to the `FILE_URI`—you get control over the image data itself. There are a number of cases where this is obviously far superior:

- You don't want the pesky user to delete their images. Silly user!
- You want to send the image to a remote server. Since there's no way to dynamically populate an `<input type="file">` tag through JavaScript, and most mobile devices don't support that input type anyway, sending the raw data through an `XmlHttpRequest` is the best bet for most PhoneGap use cases.
- Manipulating the image data: with the base-64 encoded data, it's easy to draw the given image to a `<canvas>` tag on the page and apply some transformations. This is possible using a file path also, but due to the vagaries of different paths on different platforms, it's liable to hang up with strange issues.

Thanks to the data-uri support in all modern browsers, we can duplicate all of the use cases for using a file path while also allowing us to persist the data itself.

Ensure quality is set

While there are lots of great use cases for the `DATA_URL` option, it's unlikely to be the default in PhoneGap moving forward. From an application stability perspective, sending large strings back and forth between native code and the JavaScript environment is very risky. If this is something that happens multiple times in a single session, your application could consume enough memory to crash it entirely.

The `quality` option is your friend; we have been setting quality to `50`, so it is half the quality of the source image. The quality is a value between 0 and 100; unfortunately, there's no way to specify "give me the highest quality that fits in the available memory". This seems safe for the time being; one would imagine that as devices increase their camera resolution, they also increase their available memory. However, for the nth time, we'll remind you to test all of your applications on real devices, with real use cases. If your application is seeing crashes, you can try tuning the quality down a bit further.

Pop quiz: Destination types

1. Which `destinationType` is best suited for each of the following applications?
    a. A photo-sharing application, which sends images to the user's friends
    b. A high-resolution image viewer, where users can zoom into the high-megapixel pictures taken with their camera
    c. An image manipulation application, allowing images to be stretched and squeezed with a multitouch interface

2. Why do data-uri image sources require the `data:image/png;base64,` prefix?
    a. To stop the browser interpreting them as file paths
    b. To stop the browser interpreting them as **jpeg** images
    c. To stop the browser from reloading the page
3. Why are file paths not a good choice for persistent data?
    a. The same path won't be equivalent on different devices
    b. Images can be modified or deleted without the knowledge of your application
    c. Access restrictions can prevent your application from displaying the image itself.

Editing or accessing live data

A lot of great applications that use the cameras on mobile devices don't use still images; they are used to edit videos, or to access a live video stream for augmented reality, or even just video chat. How can you do this with PhoneGap?

As of PhoneGap 1.0, a new **Media Capture** API has been added to the project. The Media Capture API is designed as a generic interface for accessing image, audio, and video data. Image data is accessed through a `captureImage` function that is very similar to the `getPicture` function; similar interfaces are available for video and audio. Like `navigator.camera.getPicture`, the capture functions call out to the native controls for recording data (audio and video, as well as images), and return the captured data back to the PhoneGap application.

Unfortunately, the state of HTML5's video and audio APIs mean there's not a whole lot you can do with the data once you receive it; you can display the data and, in the case of video, do some CSS-based manipulation, but for more complex work, you'll need to delve into the native code itself.

Summary

In this chapter, we have mastered the PhoneGap Camera API, by:

- Getting familiar with `navigator.camera.getPicture` and it's many quirks
- Grabbing file paths to draw images directly to our DOM
- Accessing raw image data for persistent storage of images

Does it seem like PhoneGap APIs are getting more complicated, and less predictable in their behavior? Are you beginning to feel the pain of mobile development pushing through our cross-platform veneer? Then you're ready for the next chapter—the contacts API!

9
Reading and Writing to Contacts

A slight etymological digression: the name **PhoneGap** *derives from the term phone, a common abbreviation used in the twentieth century to refer to the telephone, a communications device quite popular in that era. Like the watch, the calculator, or the book, the telephone was once a popular piece of technology, but has been superseded by the Internet-connected mobile computer we all carry in our pockets. A twentieth-century person would use their phone to stay in touch with their personal contacts—nowadays, we do the same thing with JavaScript.*

PhoneGap allows us to access a user's contacts programmatically, with their permission, to perform a host of actions. In this chapter, we will learn:

- How to read from the contacts: to specify a filter and retrieve every contact that matches the filter
- The differences of access and permission available on different mobile platforms: exactly how much access we will be granted to a user's contacts
- How to create new contacts that can be accessed by our own application or by others on the system

We will kick things off with an example of reading contacts from the device.

Time for action – navigator.service.contacts.find

After the last few chapters, our postcard application has gotten a little unwieldy, so we will start with a brand-new application. Create a new directory for this application, called `Find A Friend`, and we can kick things off:

1. Let's start by creating our markup and our styles for the new application. Create a file called `index.html`, with the following contents:

```
<html>
    <head>
        <title>Find A Friend</title>
        <link rel="stylesheet" href="style.css" />
        <meta name="viewport" content="width=device-width,initial-
         scale=1.0" />
    </head>
    <body>
        <h1>Find A Friend</h1>
        <input type="text" id="friendName" placeholder="What's
         your
         friend's name?" />
        <button id="friendSubmit">Submit</button>
    </body>
    <script src="phonegap.js"></script>
    <script src="xui.js"></script>
    <script src="mustache.js"></script>
    <script src="app.js"></script>
</html>
```

2. Be sure to copy `xui.js` and `mustache.js` into your new application—we will need those later on. The `phonegap.js` file will be automatically generated by each platform, and we'll start a brand-new `app.js` file shortly.

   Now create a `style.css` file with the following code:

```
body {
    background: #69F;
    font-family: Helvetica Neue, Arial;
}
h1, input, button {
    -webkit-box-shadow: #03c 4px 4px 3px;
    border-radius: 15px;
    display: block;
    margin: 8% auto;
    padding: 3% 5%;
    width: 65%;
```

```
}
h1 {
    background: blue;
    color: white;
    text-align: center;
}
input, button {
    -webkit-appearance: none;
    border: transparent;
    font-size: 0.8em;
}
button {
    background: #06C;
    color: #9CF;
    font-weight: bold;
}
button:active {
    background: #9CF;
    color: #06C;
}
```

3. Open things up in Safari to check out how they look:

The color scheme may be a little derivative, but here we are

5. Now let's write the JavaScript—here's a starting point for our `app.js`:

```
var TAP = ('ontouchend' in window) ? 'touchend' : 'click';

document.addEventListener('DOMContentLoaded', function () {
    x$('#friendSubmit').on(TAP, function () {
        var filter = x$('#friendName')[0].value;

        if (!filter) {
            // no contents
            return;
        } else {
            // name entered
            alert("we're looking for " + filter);
        }
    });
});
```

Some simple DOM stuff there—if the user has entered something to search for when touching the **Submit** button, we need to find that contact. Note the `TAP` variable at the top: conditionally binding the `touchend` event allows us to get the best performance on mobile devices while still being able to debug in our desktop browser.

6. Now we need to actually query the device's contacts. We're going to write a separate function to handle the call to `contacts.create`, since it is quite a long-winded one:

```
function findContactByName(name, callback) {
    function onError() {
        alert('Error: unable to read contacts');
    };

    var fields = ["displayName", "name"],
        options = new ContactFindOptions();

    options.filter = name;
    options.multiple = true;

    // find contacts
    navigator.service.contacts.find(fields, callback, onError,
    options);
}
```

7. Don't worry about what we need to do to make the actual call—we can think about that later on. For now, just note that we pass a name and callback function, which will execute when we get the results back.

   Here is the modified code to call `findContactByName`:

```
if (!filter) {
    // no contents
    return;
} else {
    findContactByName(filter, function (contacts) {
        alert(contacts.length + ' contact(s) found matching "' +
        filter + '"');
    });
}
```

8. Let's try this one in the iPhone Simulator first:

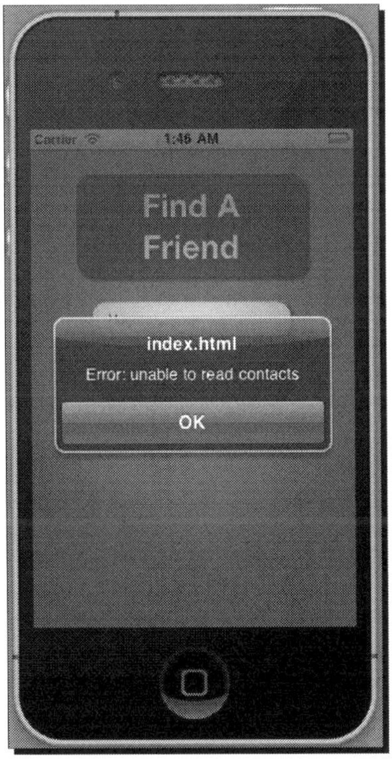

Not great: unfortunately, this is one of those cases where the simulator behavior does now map very well to that of the device.

9. If we test on an actual device, we can see some useful results; here is the expected result, as seen on my iPod Touch:

10. And for good measure, let's ensure I can find Hugo on my Android phone with the same code:

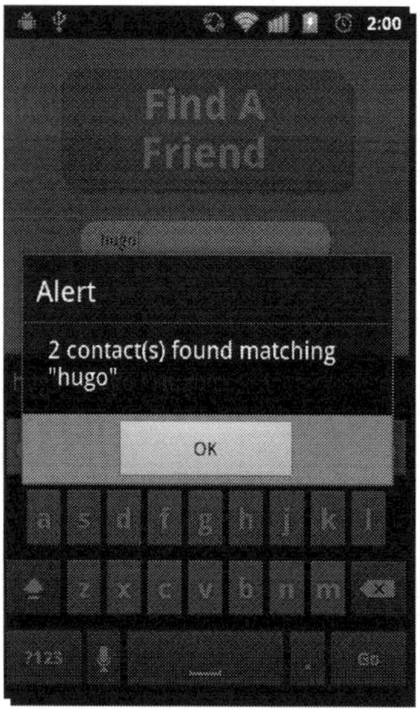

Great! Now let's do something with it.

11. As before, we're going to use the Mustache templating library to easily insert some new contents into our DOM. Let's start by adding a `<ul>` container into the DOM to contain the list of matching contacts:

```
<body>
    <h1>Find A Friend</h1>
    <input type="text" id="friendName" placeholder="What's your
     friend's name?" />
    <button id="friendSubmit">Submit</button>
    <ul id="friendsList"></ul>
</body>
```

and we'll style that list appropriately:

```
h1, input, button, ul {
    display: block;
    margin: 8% auto;
    padding: 3% 5%;
    width: 65%;
}
...
ul {
    color: #CCF;
    font-weight: bold;
}
```

12. Now we need the template for the list item itself. Add this declaration to `app.js`:

```
var item = '<li>{{ name }}</li>';
```

and let's modify the callback to contacts.find:

```
if (!filter) {
    // no contents
    return;
} else {
    findContactByName(filter, function (contacts) {
        var i = 0, contactItem, data;

        for (i; i<contacts.length; i++) {
            data = { name: contacts[i].name.formatted }
            contactItem = Mustache.to_html(item, data);
            x$('#friendsList').bottom(contactItem);
            x$('#friendsList').bottom(contacts[i].name);
        }
    });
}
```

13. Note that we need to modify the resulting contacts a bit—the objects that are returned are a little complex, so we're massaging them a little before passing to Mustache. In this case, the APIs are the same between iPhone and Android, so we just need to get the `formatted` property of the `name` property of the each contact object and pass it on. Phew.

Let's confirm that this looks good on our iPod:

And now check things on Android...

And now we're done!

What just happened?

We were able to specify a filter for the Contacts API to retrieve and access a group of contacts. We also were able to specify which fields for the contact were returned from the function call, and deal with the resulting object in order to display the contact information to the user.

Earlier releases of PhoneGap have wildly divergent Contacts APIs—the developers were essentially designing their own APIs, with the goal of providing as much of the available native functionality to PhoneGap developers as possible, with a suitably intuitive JavaScript API. As of PhoneGap 0.9.4, the API has begun to stabilize around that specified by the W3C's Device API working group (DAP)—if you're curious, this specification can be viewed at `http://dev.w3.org/2009/dap/contacts/`.

The goal of the W3C's work is to provide unified access to a single address book owned by the user, and to do so in an appropriately secure fashion. The user, according to the spec, should be allowed to freely specify and restrict the kinds of data that are exposed by the Contacts API. In my opinion, this led to a fairly obtuse and unintuitive interface for accessing the device; that said, as it's the specification that device manufacturers will likely conform to once these APIs make it into web browsers, it's worth the investment to learn.

In practice, with PhoneGap's focus on building these APIs on existing mobile platforms, such an interface can be a challenge to work around. For example, with the existing port of PhoneGap to Palm's webOS 1.x platform, the only contacts that can be read are the ones that the application has written itself—laudable from a security perspective, but next to useless for application developers. Thankfully, on the the major platforms we have been focusing on, there are pretty extensive facilities available to application developers.

One of the curious features of the DAP specification that we've seen is the **ContactFindOptions** object. A device could possibly have hundreds or thousands of contacts in its address book, and so narrowing down the quantity of results returned is essentially to maintain acceptable performance in our applications.

The ContactFindOptions object, as defined by the DAP and implemented in PhoneGap, contains three key fields:

- **filter**: A string to filter the returned results by
- **multiple**: A Boolean value that indicates whether multiple contact objects should be returned (the default is true)
- **updatedSince**: A JavaScript date object that limits the returned contact objects based on when they were last updated

At the time of writing, the `updatedSince` field is only supported on iOS devices. Typically, you'll want to use the filter option to limit the particular contacts that are returned—in most cases, you'll want multiple results, at least to give the user the choice of only selecting a single one.

ContactFields

The other unusual parameter that gets passed to `navigator.services.contacts.find` is the list of fields to be returned. The PhoneGap documentation lists 17 different fields that can be requested from the device; these range from the obvious, such as `name` and `emails`, to the imprecise, such as `note` and `categories`.

Of particular note are the fields that have the **ContactField** type, including `phoneNumbers`, `emails`, `categories`, and `urls`. Each of these fields can return an array of ContactField objects that have three properties:

- `type`: A descriptive string about this field. For example, different phone numbers for a contact could have the types **home**, **work**, and **mobile**.
- `value`: The value of this field. In the above example, the phone number for each of the three objects would be the value attribute.
- `pref`: A Boolean value that indicates which of the fields is the preferred entry. Typically, only one of all the fields would return true.

In many cases, such as `phoneNumbers`, all of these fields are set on the device's address book itself. For others, such as the `images` field, there may be unique values: in the case of `images`, the `type` property returns either **url** (for a path to the image) or **base64** (where the value property is the encoded image-data).

Ultimately, the best way to get started is to dive right in; the APIs are still in flux (and may well have changed by the time you're reading this book) and different platforms will expose different kinds of data to the user. That's the nice thing about using PhoneGap—it makes it really easy to get your hands dirty.

Have a go hero

The previous notes should have made it pretty clear that contact data can be especially heterogeneous, and dependent on the version of PhoneGap, the version of the underlying operating system, and the content a specific user will have saved to their device.

Can you modify our example code to get more useful data from your own device? How many of your contacts have associated emails or phone numbers, and how easy is it to work with those contact objects? Can you write cross-platform code that will work with richer data than we were able to use?

Writing contact data

Of course, dealing with existing contact data is only half of the fun. PhoneGap also exposes APIs for creating new contacts, and saving those to the device's global address book, as it is exposed to applications (usually inside the device's internal storage). Let's get started working with these APIs.

Time for action – Making friends

First, we can set up the view for our new **Make A Friend** action:

1. Let's modify the markup on our `index.html` appropriately:

```
<body>
      <section id="find">
        <h1>Find A Friend</h1>
        <input type="text" id="friendName" placeholder="What's
         your
         friend's name?" />
        <button id="friendSubmit">Submit</button>
        <ul id="friendsList"></ul>
        <a href="#" target="make">Make a new friend</a>
      </section>

      <section id="make" style="display: none">
        <h1>Make A Friend</h1>
        <form id="newFriend">
            <input type="text" id="newName" placeholder="What's
             your friend's name?" />
            <input type="tel" id="newPhone" placeholder="What's
             your friend's phone number?" />
            <input type="email" id="newEmail" placeholder="What's
             your friend's email?" />
        </form>
        <button id="newSubmit">Submit</button>
        <a href="#" target="find">Find an old friend</a>
      </section>
    </body>
Now style our new links appropriately (in style.css):
h1, input, button, ul, a {
    display: block;
    margin: 8% auto;
    padding: 3% 5%;
    width: 65%;
}
...
```

```
a {
    color: #fff;
    text-align: center;
    text-decoration: none;
}
```

2. Add the JavaScript to `app.js` to show and hide our new view:

```
function showSection(sect) {
    x$('section').each(function () {
        if (this.id == sect)
            x$(this).setStyle('display', 'block');
        else
            x$(this).setStyle('display', 'none');
    });
}

document.addEventListener('DOMContentLoaded', function () {
    // ...
x$('a').on(TAP, function () {
        if (this.target)
            showSection(this.target);
        return false;
    });

});
```

3. Now check all that in Safari to make sure everything works okay:

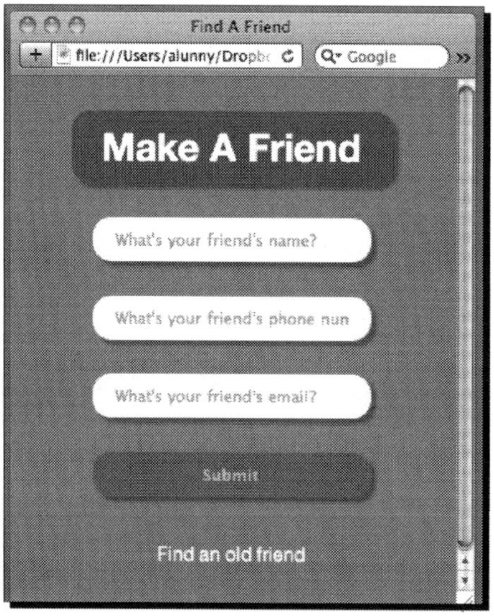

4. Now for the fun stuff—let's write a function that creates the new contact based on what the user enters. Here's a first pass:

```
function createAndSaveContact(details, callback) {
    var friend, number, email;

    try {
        friend = navigator.service.contacts.create();
        friend.displayName = details.name;
        friend.nickname = details.name;

        number = new ContactField('default', details.phone, true);
        friend.phoneNumbers = [number];

        email = new ContactField('default', details.email, true);
        friend.emails = [email];

        friend.save();

        callback.call(this, friend);
    } catch (e) {
        alert(e);
    }
}
```

Here is the verbose Contacts API in practice—lots of Java-ish new object creation, along with synchronous methods (which are somewhat of an anomaly with PhoneGap APIs).

5. Now, we just need to hook that API up to the submit button on the **Make A Friend** view—as always, make sure that this event handling code is placed in the `DOMContentLoaded` block of code:

```
x$('#newSubmit').on(TAP, function () {
    var name = x$('#newName')[0].value,
        email = x$('#newEmail')[0].value,
        phone = x$('#newPhone')[0].value;

    createAndSaveContact({
        name: name,
        email: email,
        phone: phone
    }, function (newContact) {
        alert("saved new contact " + newContact.displayName);
```

```
            showSection('find');
        });
    });
```

We can test that code in the iPhone Simulator, and it will succeed without error:

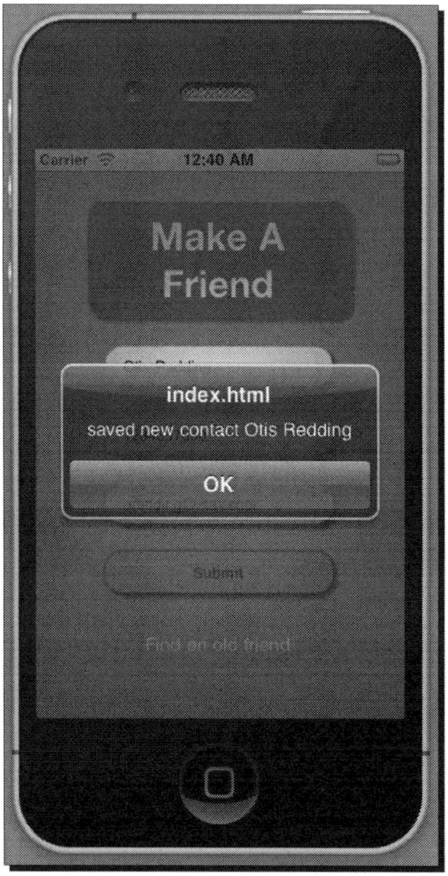

6. Unfortunately, when we try to verify the code, we're still unable to access the contacts through the simulator. We'll have to test again on a real device.

   I tried on my iPod Touch, and it did successfully save to the device's Address Book. Here's what was saved:

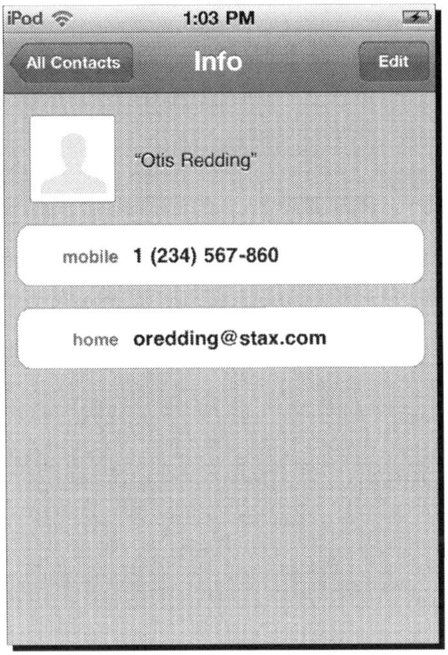

7. There are a couple of gotchas with getting the same code to work correctly on Android phone:
   - A newly added contact will not display by default unless it is associated with the user's Google Account. To fix this, go to **Display Options** on the menu for the Contacts application, expand the section with your Google Account, and ensure the **All Other Contacts** option is selected.
   - In my testing, I had to add the following permission to my Android manifest file:

     ```
     <uses-permission android:name="android.permission.GET_
     ACCOUNTS" />
     ```

   - This may just have been an anomaly—the `READ_CONTACTS` and `WRITE_CONTACTS` permissions should be sufficient.

Once both these are in place, you should find something like this:

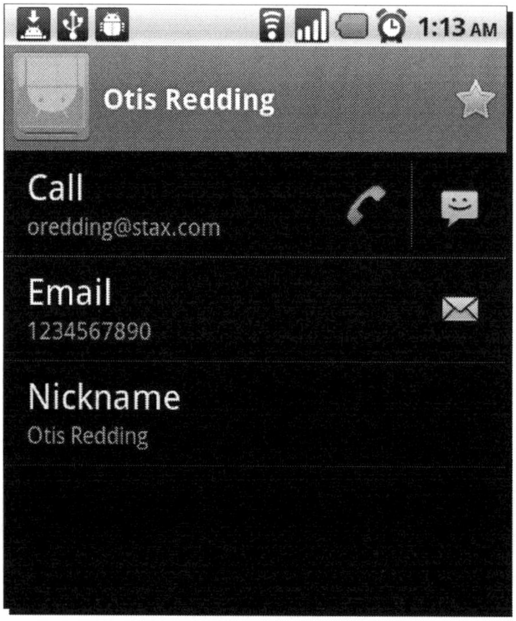

8. With some further debugging of the iOS build, an interesting gotcha has emerged:
    - The `type` field for the `ContactField` attributes can not take an arbitrary value—it's much happier with either **home**, **mobile**, or **work**

Taking account of this current quirk, my modified `createAndSaveContact` looks like this:

```
function createAndSaveContact(details, callback) {
    var friend, number, email;

    try {
        friend = navigator.service.contacts.create();
        friend.displayName = details.name;
        friend.nickname = details.name;

        number = new ContactField('home', details.phone, true);
        friend.phoneNumbers = [number];

        email = new ContactField('home', details.email, true);
        friend.emails = [email];
```

```
            friend.save();

            callback.call(this, friend);
        } catch (e) {
            alert(e);
        }
    }
```

And I even figured out how to view the Contacts application on the iPhone Simulator (press the home button, then press the **Contacts** icon) like so:

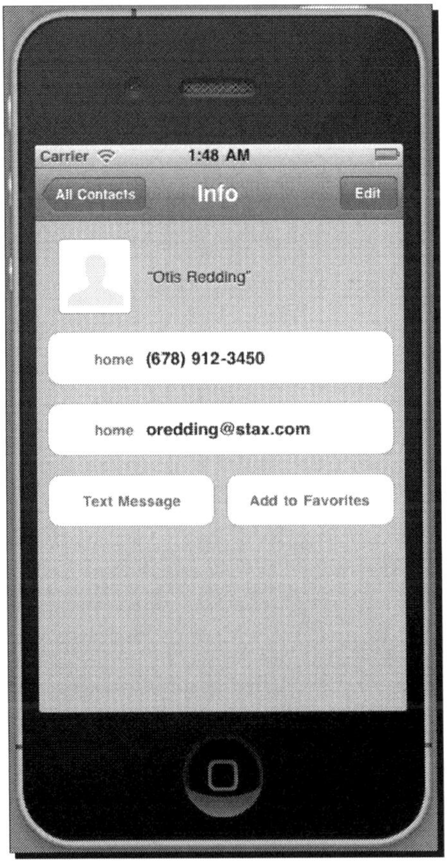

What just happened?

Well, after a great deal of amateurish hemming and hawing, we managed to write some cross-platform that successfully created a new contact on the system Address Book based on user input. That's a real feather on our caps.

I imagine it's evident by now, but I have a real distaste for the Contacts API as it stands—lots of verbosity and ceremony around a fairly simple concept. There are callbacks that we can pass to `contact.save`, but there's no easy way to perform validation on the contact objects that are passed to the device, or to know exactly what will be presented to the user without manually testing on individual devices. Unfortunately, that's simply the nature of this particular beast—since contact management is such an intrinsic part of modern smartphones, it's no surprise that the individual implementations differ so wildly.

What if I encounter a new problem?

Because of the peculiarly difficult problems of the Contacts API—honestly, it never does what I want it to—you may well encounter new roadblocks that have not been covered in this chapter.

The two best resources are the PhoneGap mailing list, hosted on Google Groups and easily found, and the PhoneGap documentation website, available at `http://docs.phonegap.com`. The documentation website gives the best listing of the various API calls available, which parameters are expected for which calls, and what arguments are passed to your callback variables. In the case of the Contacts API, with so must exhaustive detail to work through, it's an invaluable resource.

The mailing list, with a few thousand active PhoneGap developers, is just as useful. In particular, the mailing list is helpful for one-off or device-specific problems that you may encounter. The Android problem I saw above—where the contacts were definitely being written to the device, but I could not see them without changing my Contacts settings—was detailed on the mailing list, and your problem may well be too.

ContactFields, ContactName, and similar objects

In all of the examples above, I've used PhoneGap's JavaScript constructors to create the contact object—`new Contact` and `new ContactField`, and so forth. This is a fairly common idiom in statically typed languages, such as Java, but many JavaScript developers will be more familiar with a loose, essentially duck-typing approach.

Your PhoneGap code is of course just JavaScript, so you can use the facilities of the language: creating lightweight hashes rather than using the `new` operator, leaving optional arguments out of function calls, and defining anonymous inline functions for callbacks. The main benefit of using these constructors for the Contacts API is that they pre-populate (with null values) all of the myriad fields that will be expected by the native code. If you don't do this, unexpected behavior can arise.

The safest bet, especially if you're not comfortable digging through the PhoneGap source code yourself, is to use the tools that the framework makes available, for the most predictable outcomes.

Be responsible

PhoneGap gives fairly unfettered read and write access to a user's address book, and it's very easy to abuse that access. It should go without saying, but if you're aiming to build a valuable application for your own company or for your clients, you should be very careful with any contacts data you access.

In particular, make a point of not saving any data or sending it back to a remote server without the explicit permission of the user, don't send any communications to any of the user's contacts (again, unless the user accedes explicitly), and don't write new contacts indiscriminately into the address book. It would be a good way to get your application banished from any App Stores or App Markets in which it was listed, which will be a good way to ensure that nobody uses the application at all.

Pop quiz: Contacts

1. Which fields need to be set before `contact.save()` can be called?
    a. `emails`, `phoneNumbers`, and `name`
    b. `name`, `displayName`, and `nickname`
    c. None, strictly speaking
2. Why is it a good idea to set a filter when calling `contacts.find()`?
    a. Retrieving too many contacts could crash your application
    b. To prevent all of the contact fields from being returned
    c. To comply with the device's permissions rules
3. Why are `phoneNumbers` and `emails` set as arrays, rather than string fields?
    a. To separate the field's role from its value
    b. A contact can have multiple phone numbers and email addresses
    c. Because that's what the W3C's DAP group suggests

Summary

With much struggle, we've gotten the hang of the PhoneGap Contacts API in this chapter, in order to:

- Find contacts on the device's address book, and display information about them to the user
- Create new contacts, and persist them back to the address book
- Understand the various contact objects and interfaces, to better make full use of them

Had enough fun yet? If the Contacts API seemed like a bit of a cross-browser struggle, you ain't seen nothing yet. Yes my friend, the next chapter is full of unrestrained, no holds-barred native code, in its entire majestic, uh, majesty. That's right, it's PhoneGap Plugins!

10
PhoneGap Plugins

Over the past few chapters, we've discussed the core PhoneGap APIs—a set of JavaScript bridges allowing access to native functionality on mobile devices. While PhoneGap aims to provide an extensive set of APIs to support the most common use cases on mobile devices—accessing device sensors and user data—there's no way it can provide an exhaustive, one-to-one mapping for every device API. The goal is really to provide a cross-section of the most common native functionality that will be supported on the major mobile platforms.

When other APIs are required for your application, PhoneGap provides robust plugin architecture, allowing developers to easily add new functionality to their applications and integrate it with their existing PhoneGap applications. In this chapter, we will see how to:

- Integrate an existing PhoneGap plugin into our application, to extend its functionality
- Write a quick battery status plugin for iOS, allowing us to see the battery level for our device
- Port our battery plugin to iOS and Android, using the same JavaScript API to access the information on all devices

Let's kick things off by integrating an existing plugin.

Getting PhoneGap plugins

At the time of writing, the canonical place to get PhoneGap plugins is a GitHub repository—`http://github.com/phonegap/phonegap-plugins`. If you go to that URL, you'll see a directory structure like the following:

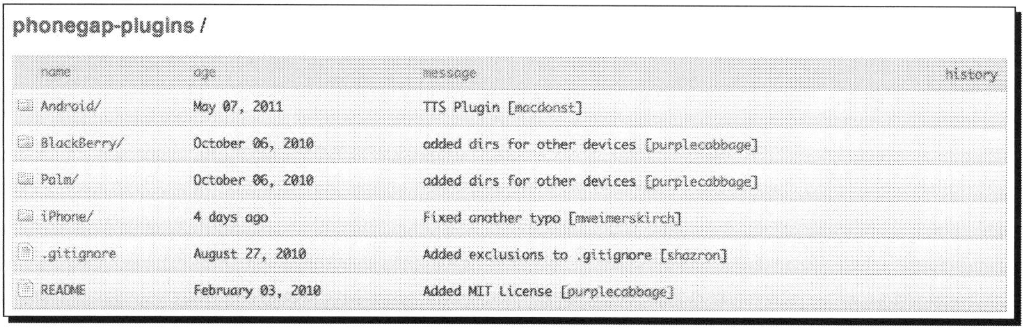
phonegap-plugins /

name	age	message	history
Android/	May 07, 2011	TTS Plugin [macdonst]	
BlackBerry/	October 06, 2010	added dirs for other devices [purplecabbage]	
Palm/	October 06, 2010	added dirs for other devices [purplecabbage]	
iPhone/	4 days ago	Fixed another typo [mweimerskirch]	
.gitignore	August 27, 2010	Added exclusions to .gitignore [shazron]	
README	February 03, 2010	Added MIT License [purplecabbage]	

The first important thing you should know about PhoneGap plugins is that they have to be implemented separately for each platform they support. Although, in the best case, each of the platforms has the same JavaScript API, the native code cannot be shared.

Use the following command to get a local copy of PhoneGap plugins on your system:

```
git clone https://github.com/phonegap/phonegap-plugins.git
```

The plugin we'll integrate is one that has been implemented across different platforms—the **ChildBrowser** plugin, that allows us to display external websites in the context of our application.

Time for action – Integrating ChildBrowser

1. Get your **Find A Friend** application that we built in *Chapter 9*. We're going to extend that by adding a link to the bottom of the page which explains what a friend is.

   Add this link to the bottom of your `body` tag, just below the last `section`:

   ```
   <a href="http://en.wikipedia.org/wiki/Friendship">What is this
   friendship you speak of?</a>
   ```

2. Make sure this code is in a PhoneGap iPhone project, and run the resulting code in the iPhone simulator. Click on the link, and see what happens:

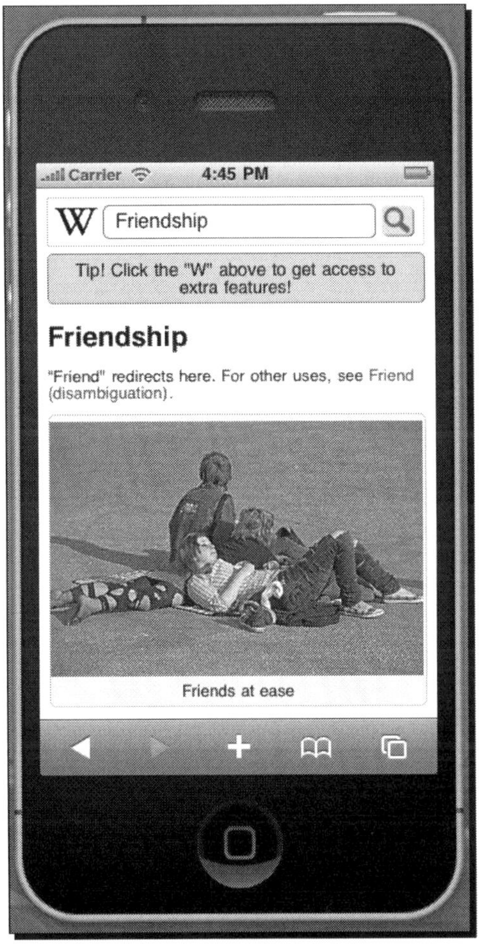

3. As we can see, the user has been kicked out of our application and Mobile Safari has been opened. This is not the ideal user experience—the ChildBrowser plugin aims to accommodate for this.

4. The ChildBrowser plugin for iOS, written by Jesse MacFadyen, is located under your clone of phonegap-plugins at `phonegap-plugins/iPhone/ChildBrowser`. It includes an iPhone view controller (a user interface class) and a **PGPlugin** subclass—we'll cover later on how PGPlugin works. There are also some resources for the view display, and a JavaScript file.

Due to how Xcode works, we'll have to drag the files into our Xcode project to ensure that they're added to the project:

To ensure the import works correctly, check that the files are copied into your project, and that the appropriate references are created in your project:

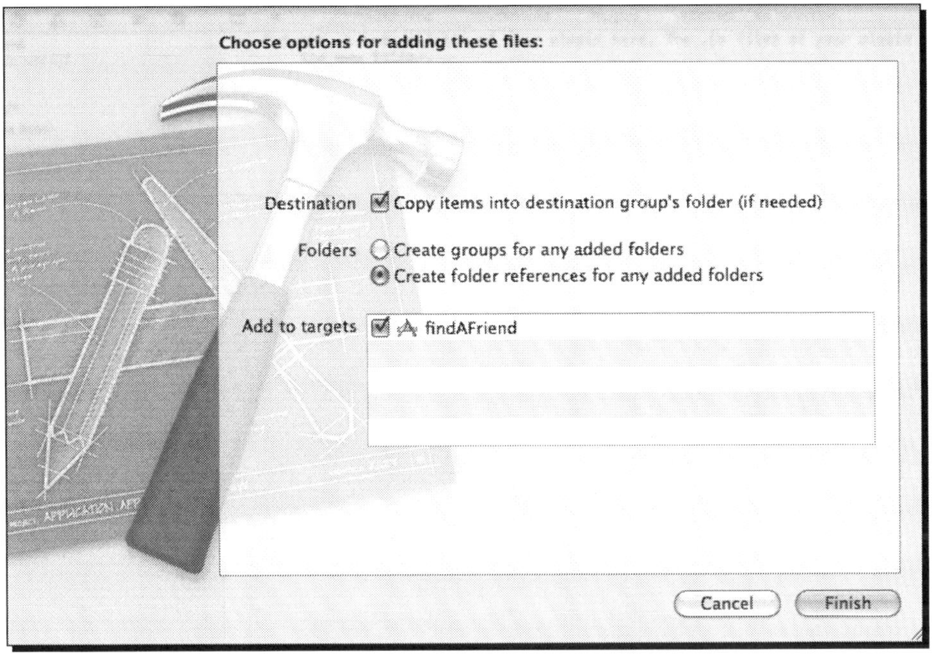

5. As of PhoneGap 1.0, you will also need to edit the `PhoneGap.plist` file, under the **Supporting Files** group of your project in Xcode. Find the **Plugins** dictionary and add a new row, with the key **ChildBrowserCommand** and the value **ChildBrowserCommand**.

   It's probably a good idea at this point to do a quick build in Xcode, to ensure that everything has been imported correctly.

   You can then copy the `ChildBrowser.js` file directly into your www directory, and modify `index.html` to ensure it's included in the page:

```
<script src="phonegap.js"></script>
<script src="ChildBrowser.js"></script>
<script src="xui.js"></script>
```

6. ChildBrowser is written to not initialize the JavaScript object automatically; we will need to initialize the plugin ourselves to access it in our application.

   Add the following code to the bottom of `app.js`:

```
document.addEventListener('deviceready', function () {
    ChildBrowser.install();
}, false);
```

And then let's modify our link handler code once more to call ChildBrowser, if we have an external link:

```
x$('a').on(TAP, function () {
    if (this.target) {
        showSection(this.target);
        return false;
    } else if (/^http/.test(this.href)) {
        if (window.plugins && window.plugins.childBrowser)
            window.plugins.childBrowser.showWebPage(this.href);
        return false;
    }
});
```

7. Make sure all the latest code is in our Xcode project, clean and build again, and we should see the link load in a child browser:

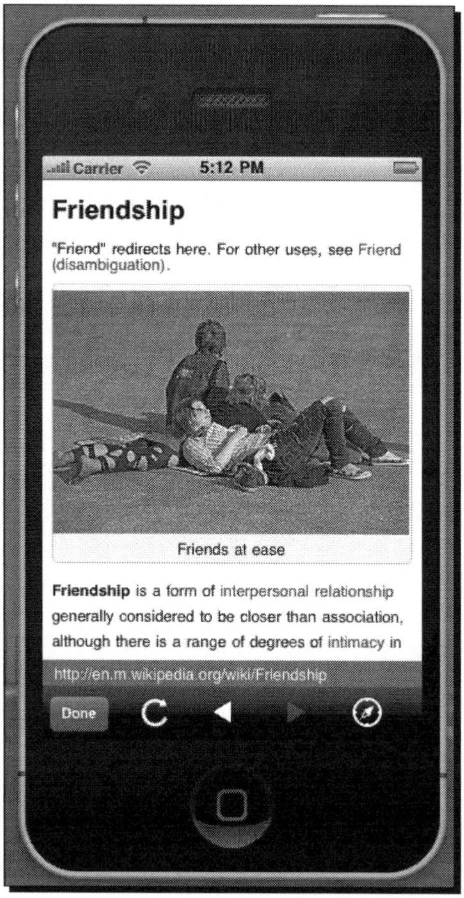

8. That's all well and good, but how can we get the same experience on Android? There is indeed an Android port of ChildBrowser, written by Bryce Curtis and located at `phonegap-plugins/Android/ChildBrowser`.

   There is a bit more ceremony involved in installing an Android plugin, but on the JavaScript side, we don't have to do much differently. Copy the Android `childbrowser.js` file into the `assets/www` directory of your PhoneGap Android project. I've renamed the file in my project to `ChildBrowser.js`, so I don't need to change the `script` tag in my `index.html` file. The Android plugin does not require a call to `ChildBrowser.install`, however, we'll add an `if` statement to prevent that call from throwing an error:

   ```
   document.addEventListener('deviceready', function () {
       if (window.ChildBrowser && ChildBrowser.install)
           ChildBrowser.install();
   }, false);
   ```

9. Now we need to add the Java ChildBrowser class into our PhoneGap Android project. Create a sub-directory in your project called `src/com/phonegap/plugins/childBrowser` and copy `ChildBrowser.java` into there.

10. The final steps are to register the plugin. Add the ChildBrowser activity to your `AndroidManifest.xml` file, to ensure that the OS knows about the new plugin—the details of this are listed in the plugin's `README` file. Then add the plugin's key to your app's `res/xml/plugins.xml` file, as follows:

    ```
    <plugin name="ChildBrowser" value="com.phonegap.plugins.
    childBrowser.ChildBrowser"/>
    ```

Now you can load your application in the Android simulator, and check out the link:

11. The Android operating system's structure of activities and intents means that the ChildBrowser plugin is arguably less useful than on iOS, but it's a good example of how to integrate similar plugins in a cross-platform fashion. It also demonstrates how we can import native code into our PhoneGap project just by moving around a few files.

What just happened?

We saw the basic mechanism for importing plugins into our PhoneGap applications, on both iOS and Android. Most of the time was spent importing and integrating the native code into our platform-specific projects; because of PhoneGap's structure, adding the JavaScript code to our applications is relatively easy.

Differences between platforms

There's no reason why there should be any difference in the JavaScript APIs between platforms, but the ChildBrowser code is a good example of why there may be differences after all. Jesse originally implemented the code for iPhone, and it reflects his own design choices: the file is called `ChildBrowser.js` and requires a call to `ChildBrowser.install`. It was ported independently to Android by Bryce, who named his file `childbrowser.js` and installed the plugin onto the `window` object by default. Although we didn't cover it above, the iOS implementation is also a lot more feature-rich than the Android one, which only provides the `showWebPage` functionality.

The differing user experiences on the two platforms also reflect the differing prejudices and best practices among mobile platforms. The iOS ChildBrowser pulls itself up within the context of the application's main view, and has its own custom chrome for displaying web pages. In contrast, the Android port sends an Android intent containing the web URL to the operating system, and allows the OS browser to handle displaying the page. Since Android handsets have a back button built into the device, returning to the source application is far easier than on iOS.

These differences are just the hazards of open source software—since there are no design documents mandating the architecture of plugins down to the individual code monkeys, different developers are free. Most developers are focused primarily on one platform, and so the plugins they write are tailored to that platform.

Plugin discovery

A corollary to the above points is that, at the time of writing, discovery for PhoneGap plugins is a bit of a mixed bag. The GitHub repository listed above is a curate collection of plugins, mostly contributed by developers on the PhoneGap project themselves, but is far from an exhaustive collection of everything that has been done, let alone all that will be done.

If you're interested in a plugin for a specific purpose, your best bet is to ask on the PhoneGap mailing list to see if any other developers have attempted to solve the same problem. If so, they may not have published or publicized their code as such, but they may be willing to send some over email. As the PhoneGap project matures past a 1.0 release, the team will be focusing on improving this situation, so that code can be more easily shared by PhoneGap developers.

Of course, you could also write a plugin yourself—and that's what we're going to cover next.

Pop Quiz: Using PhoneGap plugins

1. What best describes the process of adding a PhoneGap plugin to your application?
    a. Run the plugin's installer, pointing it towards the application you want to enhance
    b. Read the `README` for the individual plugin and follow the instructions to the letter
    c. Add the JavaScript code to your `www` directory, and add the native code to each of your individual PhoneGap projects
2. Why does the ChildBrowser plugin behave differently on iOS, compared to Android?
    a. Each one caters to the user expectation of the specific platform
    b. The two ports were developed independently of each other
    c. The iPhone developer was more intelligent than the Android developer
3. PhoneGap plugins require native code to be written independently for each platform. Why is this?
    a. Different platforms use different programming languages for their native APIs
    b. Each platform has its own particular style of API, regardless of the programming language
    c. Each platform has its own namespace, and all code that executes on the platform must be included in that namespace

Writing a PhoneGap plugin

Existing PhoneGap plugins are great but, unfortunately, you're limited to whatever extra code other developers have implemented. While the amount of open source and widely available PhoneGap plugins is growing all of the time, the real power of PhoneGap plugins is the ease with which you can add your own interfaces into your PhoneGap applications by writing a little native code.

This is the point where the skeptical reader will throw this book across the room, cursing its duplicitous author once and for all. Write native code? That's exactly what I didn't want to do! If I wanted to write native code, why would I be using PhoneGap?

Hopefully the skeptical reader has picked the book back up, regained her composure, and decided to hear me out. PhoneGap plugins are powerful because you can write lightweight bridges to native APIs, exposing the data you're interested in to the JavaScript level. Writing a couple dozen lines of native code is nothing anybody should be afraid code; it's certainly a far cry from developing your entire application against these native APIs. And the PhoneGap framework provides the plugin architecture that allows these access hooks to be quickly and effectively implemented.

We're going to look at how this works by implementing a lightweight battery status plugin for iOS; later, we'll port it over to Android, and to BlackBerry WebWorks.

Time for action – Battery view

By this point, you should be well familiar with the setup for creating a new PhoneGap application:

1. Let's create an application called **BatteryStatus**, and fill in the `index.html` as follows:

```
<!DOCTYPE html>
<html>
    <head>
        <meta http-equiv="content-type" content="text/html;
         charset=UTF-8" />
        <meta name="viewport" content="width=device-width,
        initial-scale=1.0, maximum-scale=1.0, user-scalable=no;" />
        <title>Battery Plugin Example</title>

        <link rel="stylesheet" href="style/app.css" />

        <script src="phonegap.js"></script>
    </head>
    <body>
      <div id="battery">
        <img src="style/battery.png" />
        <div id="capacity" class="full"></div>
      </div>
    </body>
</html>
```

Hopefully that's all clear—make sure you get the `battery.png` file (and other image assets) from the code samples available with the book.

Fill out the `style/app.css` file also:

```
body {
  background-color:#000000;
}
#battery {
  height:182px;
  margin:0px auto;
  width:341px;
  position:relative;
}
img {
  left:0px;
  position:absolute;
  top:0px;
  z-index:2;
}
#capacity {
  height:135px;
  left:28px;
  position:absolute;
  top:22px;
  width:270px;
  z-index:1;
}
#capacity.full {
  background-color:#2AAF2C;
}
#capacity.low {
  background-color:#af2a2a;
  width:40px;
}
h1 {
  font-family:verdana, helvetica;
  font-size:58px;
  color:#1B1D20;
  text-align:center;
  text-shadow: #888 0px 1px 0px;
}
```

2. There's a bit of busy work here to get everything precise; essentially, what we want is to display the same battery image that is shown on the iPhone lock screen, and dynamically update it with the status of the device's battery.

A quick check in Safari reveals that the full class is being applied correctly:

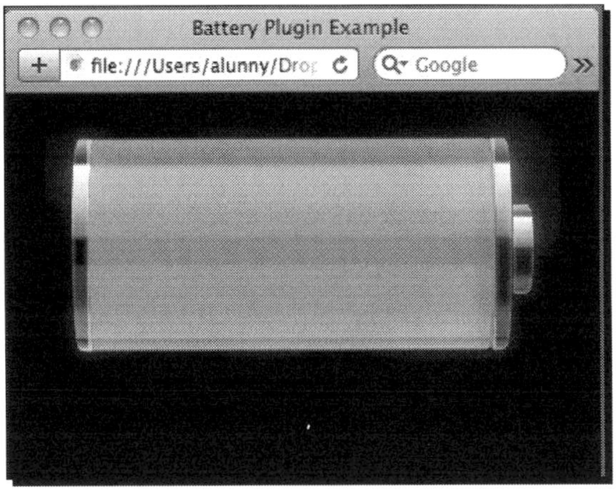

3. So far, so good. The next important step is to set up the code to modify the display of the battery itself.

Firstly, we're going to write a helper file called `color.js`, which is able to map between HSV (hue, saturation, and value) and RGB (red, green, and blue) colors, along with hex values. The two functions we're using are open source ones—I've included the URLs for their original sources:

```
(function(scope) {
    //
    // Adapted from: http://mjijackson.com/2008/02/rgb-to-hsl-and-
    rgb-to-hsv-color-model-conversion-algorithms-in-javascript
    //
    window.hsvToRgb = function(h, s, v){
        var r, g, b;

        h = h / 360;
        s = s / 100;
        v = v / 100;

        var i = Math.floor(h * 6);
        var f = h * 6 - i;
        var p = v * (1 - s);
        var q = v * (1 - f * s);
        var t = v * (1 - (1 - f) * s);

        switch(i % 6){
            case 0: r = v, g = t, b = p; break;
            case 1: r = q, g = v, b = p; break;
            case 2: r = p, g = v, b = t; break;
```

```
            case 3: r = p, g = q, b = v; break;
            case 4: r = t, g = p, b = v; break;
            case 5: r = v, g = p, b = q; break;
        }

        return [r * 255, g * 255, b * 255];
    }

    //
    // Adapted from:
    http://haacked.com/archive/2009/12/29/convert-rgb-to-hex.aspx
    //
    window.colorToHex = function(r, g, b) {
        var red   = parseInt(r);
        var green = parseInt(g);
        var blue  = parseInt(b);

        var rgb = blue | (green << 8) | (red << 16);
        return '#' + rgb.toString(16);
    };
})(window);
```

4. Next, we're going to create an `app.js` file that defines a `BatteryStatus` object, which uses these color methods to set the appearance of the battery image automatically:

```
function BatteryStatus () {
    // Max pixel width of the battery's capacity
    var max = '270';

    // Generate a color based on the charge percentage.
    var chargeColor = function(percent) {
        // Color of a fully charged batter (green)
        var hsv = { h: 121, s: 76, v: 69 };

        // Scale the hue based and convert HSV to Hex
        var h = hsv.h * (percent / 100);
        rgb   = hsvToRgb(h, hsv.s, hsv.v);
        hex   = colorToHex(rgb[0], rgb[1], rgb[2]);

        return hex;
    };

    // Set the battery charge level
    this.setCharge = function (value) {
        var el = document.getElementById('capacity');
```

```
        el.style.width = (max * (value / 100)) + 'px';
        el.style.backgroundColor = chargeColor(value);
    }
};
```

Add script tags for `color.js` and `app.js` to your `index.html` file, and test things out in Safari—you can use the Web Inspector, as shown in the following screenshot, to test that it works:

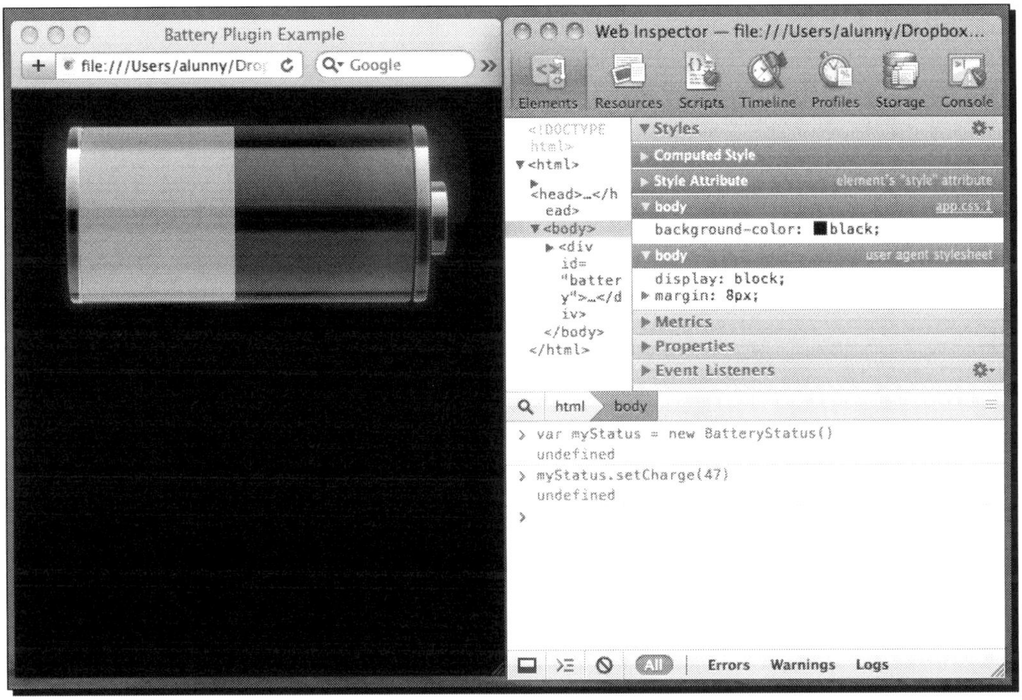

5. Now let's define an interface for our JavaScript calls to the plugin. We execute the `PhoneGap.exec` function to call into the native code, in order to get a battery level, which we'll define as being between 0 and 100. The actual code for doing this is quite succinct; we'll write a `plugin/battery.js` file that has the following contents:

```
var Battery = function() {
    return {
        get: function(property, successCallback, errorCallback) {
            PhoneGap.exec(successCallback, errorCallback,
            'Battery', 'get', [ property ]);
        }
    }
};
```

```
PhoneGap.addConstructor(function() {
    if (!window.plugins) window.plugins = {};
    window.plugins.battery = new Battery();
});
```

Hopefully that's all clear—it added `window.plugins.battery` as our battery plugin object.

Ensure that this code gets loaded by inserting the correct script tag into your `index.html`:

```
<script src="phonegap.js"><script>
<script src="plugins/battery.js"><script>
```

6. Let's add another function to our `BatteryStatus` object in `app.js` to keep polling `window.plugins.battery` for the latest status:

```
function BatteryStatus () {
    ...
    // Monitor the battery charge level... forever.
    this.watchCharge = function() {
        var self = this;

        window.plugins.battery.get(
            'Power',
            function(data) {
                if (!data.level)
                    data = JSON.parse(data);
                self.setCharge(data.level);
                setTimeout(function() { self.watchCharge(); },
                100);

            },
            function(e) {
                alert('battery watch error: ' + e);
            }
        );
    }
};
```

The `data = JSON.parse(data)` is to work around the fact that we're passing strings, not JavaScript objects as such. It's a little ugly, but it ensures our code will work cross-platform, once we get there.

7. Now for the fun part. Make sure you have an Xcode project set up for your application, and ensure that all of the code that we've written is in place.

Go into Xcode, right-click on the Projects group, and select **New File...**. Create a new Objective-C class called `Battery`.

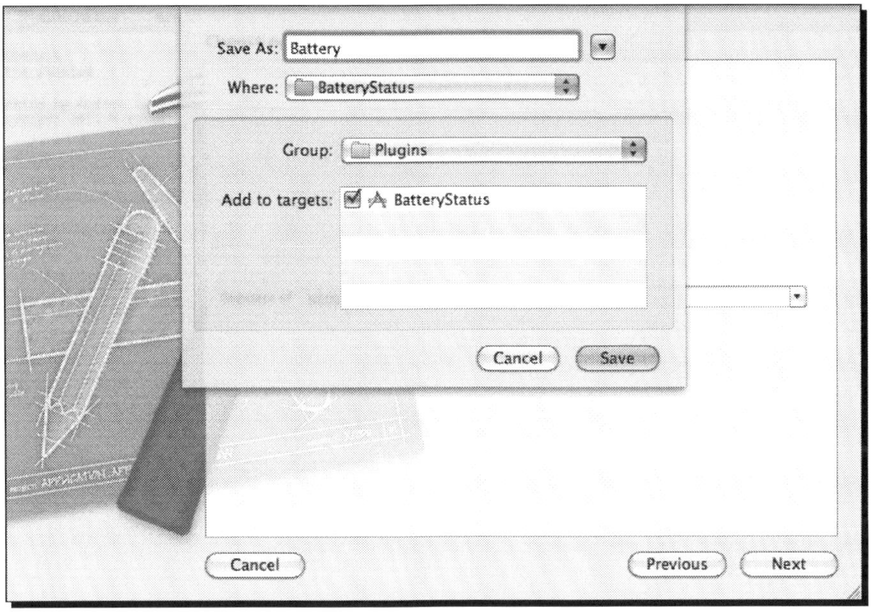

In the header file—`Battery.h`—make sure that your class inherits from `PGPlugin`, and has one method defined—`get`. Here's what we'll enter:

```
#import <Foundation/Foundation.h>
#import <PhoneGap/PGPlugin.h>

@interface Battery : PGPlugin {
    NSString *win;
    NSString *fail;
}

-(void) get:(NSMutableArray*)arguments
        withDict:(NSMutableDictionary*)options;

@end
```

Creating two instance variables for our success and failure callbacks allows us to easily cache the callbacks we're targeting, in case we have multiple, asynchronous function calls on the native side before we return any content to the web view.

We also need to register our plugin in our `plugins.plist` file. We will do this as we did with the ChildBrowser plugin—by adding a single row to the **Plugins** dictionary with the same key and value (in this case, **Battery**).

8. The next point is to get our code calling back into the web view—without worrying about the API for getting the battery data, let's just send some value back over the bridge. Here's the method we're going to put into `Battery.m`—read it over, and I'll explain afterwards:

```
-(void) get:(NSMutableArray*)arguments
   withDict:(NSMutableDictionary*)options {
    win = [arguments objectAtIndex:0];
    fail = [arguments objectAtIndex:1];
    NSString* jsString = NULL;
    PluginResult* result = nil;

    @try {
        NSUInteger status = 50;

        result = [PluginResult resultWithStatus:PGCommandStatus_OK
                                messageAsString:[NSString
                                stringWithFormat:
                                @"{\\\"level\\\":%d}", status]];
        jsString = [result toSuccessCallbackString:win];
    }
    @catch (NSException *exception) {
        result = [PluginResult
        resultWithStatus:PGCommandStatus_ERROR
        messageAsString:@"error: could not read battery!"];
        jsString = [result toErrorCallbackString:fail];
    }
    @finally {
        [[self webView]
        stringByEvaluatingJavaScriptFromString:jsString];
    }
}
```

Bizarre Objective-C syntax aside, here's what is happening: we read the first two arguments as the success and error callbacks (`win` and `fail`). We attempt to get a plugin result callback string including the current status (hard-coded to 50). If that fails, we call the error callback using a stock error message. Finally, we call whatever JavaScript string we ended up with.

Confused? Me too, but here's how it looks in the simulator:

9. Finally, actually retrieving the battery status.

   iOS exposes the battery level as a property on the current device (`UIDevice currentDevice`, in Cocoa parlance). We can get it quite easily, but there are a couple of gotchas:

   - The `batteryLevel` property is a float between 0 and 1, and we want an integer between 0 and integer. An easy enough fix, you'll agree.
   - If the level is unknown (as on the simulator), -1.0 is returned as the level. In this case, we'll return a default value of 42.

   Here is the code to get the battery level, in the format we want:

```
float rawStatus;
NSUInteger status;

rawStatus = [[UIDevice currentDevice] batteryLevel];
```

```
if (rawStatus < 0)
    status = 42;
else
    status = rawStatus * 100.0;
```

Of course, remove the other assignment of status to 50, and you should be done. Exciting battery levels abound!

What just happened?

We wrote our first iOS PhoneGap plugin—a simple one to read the device's current battery level and display it to the user, in a graphically pleasing manner.

For a long time, PhoneGap plugins on iOS were an ad hoc affair. The iOS implementation of PhoneGap is based on the Cocoa Touch `UIWebView` class that has a mealy-mouthed method called `stringByEvaluatingJavaScriptFromString` to directly execute JavaScript code in the context of the webview. We could quite easily roll our own callback mechanism on the JavaScript side, and cut the Objective-C code down to building a JavaScript string and calling it directly.

However, a lot of the subsequent work on standardizing plugins across the different plugins has paid dividends for providing a standard interface for communication between native code and JavaScript code. In particular, the `PluginResult` class encapsulates the ceremony of storing references to the two callbacks (success and error) while allowing an easy method of communicating the status of each of the native operations. Our Battery method was able to easily map the overall structure of one of the standard PhoneGap APIs, without much boilerplate code to achieve that aim.

Noteworthy information about the PhoneGap plugin with iOS

- Objective-C, the language of native iOS development, has a few nice object-orientated features that facilitate easy plugin development. One of them is dynamic object creation and method lookup, meaning we can get an instance of an object and call one of its methods just by knowing the names of the class and the method as strings. We can thus avoid a lot of ceremony around registering native objects and methods, and quickly get something up and running.
- iOS, or Cocoa Touch, is very flexible when it comes to views, or native components of views. If there is a native user interface library you're interested in hooking up with your application—for example, Facebook's **Three20** library, which provides a fully featured and performant image gallery—a little understanding of iOS views and view controllers will allow you to show and hide a modal display within the context of your PhoneGap application, as the ChildBrowser plugin does.

- Communication between native code and web code on iOS is all done through strings, which puts practical limits on how much data can be transmitted at once. If you're aiming to move very large amounts of data across that bridge, you may have to look at buffering or throttling the communications somehow.

The most important thing our demo should have shown is how little native code you actually need to write to implement a PhoneGap plugin—I had to look up a single method call, to access the `batteryLevel` property of the device in Cocoa Touch, and then it was just using the PhoneGap Plugin Result API to communicate back to the JavaScript side. This is good to keep in mind—writing a PhoneGap plugin does not mean learning a completely new development environment.

Have a go hero

The battery level was pretty good, but are there any similar quick wins you can get from a cursory glance at the iOS documentation? What other methods are available on the `UIDevice currentDevice` object? Can you write a PhoneGap plugin to determine if the user is physically close to their device?

Porting your plugin

Writing a plugin for one platform is all well and good, but that's not why we're PhoneGap developers. We want to have the same code run in separate environments, with minimal changes, ideally. Luckily, we can do that quite easily, although for this particular example the other platforms we're targeting do take a little more code. But who's worried about that; let's dive right in!

Time for action – Android and BlackBerry

Android first—get your Android project set up through Droidgap as usual. You want to have a bare PhoneGap Android project ready, with the contents of the www directory from our iPhone example.

1. We will need to make one small change to our JavaScript, in `plugins/battery.js`. As mentioned above, Objective-C has all sorts of metaprogramming and reflective goodness to ensure we don't need to explicitly register our plugins. Unfortunately, Android, a Java-based platform, has no such niceties.

   We need to edit our `addConstructor` call appropriately:

```
PhoneGap.addConstructor(function() {
    // add plugin to window.plugins
    if (!window.plugins) window.plugins = {};
    window.plugins.battery = new Battery();
```

```
    // register plugin on native side
    if (navigator.app && navigator.app.addService)
        navigator.app.addService('Battery', 'com.phonegap.plugins.
Battery');
});
```

I've coded especially defensively in these examples to ensure cross platform compatibility, but all PhoneGap versions 0.9.5 and above should expose `navigator.app.addService`. Still, it's always worth treading with caution.

2. Android doesn't expose a single property for the battery level, so we're going to create two separate classes for this purpose. Firstly, create the directory `src/com/phonegap/plugins` in your Android project.

   The first class we'll create in that directory will be called `BatteryReceiver.java`. Here's the code:

```
package com.phonegap.plugins;

import android.content.BroadcastReceiver;
import android.content.Context;
import android.content.Intent;
import android.os.BatteryManager;

public class BatteryReceiver extends BroadcastReceiver {
    private int batteryLevel = 0;

    public int getLevel() {
        return batteryLevel;
    }

  @Override
    public void onReceive(Context arg0, Intent intent) {
    batteryLevel = intent.getIntExtra(BatteryManager.EXTRA_LEVEL,
0);
  }
}
```

   `BatteryReceiver` is a subclass of `BroadcastReceiver`; it's a class that receives messages from the operating system. Android will periodically tell us what the battery level of the device is—often enough it's not noticeably slower than a property access on iOS. Each time we receive one of these messages, we're going to cache that as an instance variable.

3. This is all well and good, but it's not communicating with the web view at all. The other class we'll create is called `Battery`, just like the iOS `PGPlugin` subclass. `Battery` inherits from `com.phonegap.api.Plugin`, and looks like so:

```
package com.phonegap.plugins;

import org.json.*;
import com.phonegap.api.PhonegapActivity;
import com.phonegap.api.Plugin;
import com.phonegap.api.PluginResult;

import android.content.BroadcastReceiver;
import android.content.Context;
import android.content.Intent;
import android.content.IntentFilter;
import android.os.BatteryManager;

public class Battery extends Plugin {
    private BatteryReceiver bReceiver;

    public void setContext(PhonegapActivity ctx) {
        super.setContext(ctx);
        IntentFilter batteryLevelFilter = new
        IntentFilter(Intent.ACTION_BATTERY_CHANGED);
        bReceiver = new BatteryReceiver();
        ctx.registerReceiver(bReceiver, batteryLevelFilter);
    }

    public PluginResult execute(String action, JSONArray args,
    String callinglbackId) {
        int level = bReceiver.getLevel();
        try {
            return new PluginResult(PluginResult.Status.OK,
            "{\"level\":" + level + "}");
        } catch(Exception e) {
            return new
            PluginResult(PluginResult.Status.INVALID_ACTION,
            "error: could not read battery!");
        }
    }
}
```

It's a lot more verbose than the iOS version (perhaps because I'm less well-versed in the Android APIs), and it's doing a few more interesting things.

The plugin interface is much the same as on iOS. The Android Plugin interface, however, has a method you can override called `setContext`, which essentially acts as a constructor, with a link back to the main `PhoneGapActivity` object (that is, the web view, and associated context).

In our `setContext` method, we initialize a `BatteryReceiver` object, and assign it to listen for Intents (messages to you and I) about the battery level. All intents that pass through that filter are received by the `BatteryReceiver`.

All calls to the plugin from JavaScript pass through the `execute` method. This is a crucial difference from iOS: instead of JavaScript APIs having access to any method on your class, all plugin calls on Android go through a single method, with the exact call (the JavaScript function, usually), passed as the first parameter, `action`. In this case, we only have a single action we're interested in, so we ignore that parameter. We just grab the most recent battery level reading from our `BatteryReceiver` and return it back as a `PluginResult` object. PhoneGap Android handles the rest.

Finally, as with ChildBrowser, we need to register the battery plugin in our `res/xml/plugins.xml` file:

```
<plugin name="Battery" value="com. phonegap.plugins.Battery" />
```

There's a lot going on there, but if all goes well, the resulting user experience should be much the same as on iOS:

4. And so finally, we come to BlackBerry WebWorks. Thankfully, on BlackBerry 6.0 and above devices, we won't have to make any changes to our JavaScript code. Phew!

   The native Java code for BlackBerry WebWorks is very similar to that on Android—a `Plugin` class to inherit from and a `PluginResult` to return from methods. The subclass of Plugin has an important `execute` method, that is passed an initial parameter, `action`, that tells the plugin which action to perform.

   I'm indebted to my coworker Michael Brooks for his implementation of the Battery level plugin on BlackBerry (and doing much of the styling for view of the app). Like Android, there are two classes for the BlackBerry implementation: a `Battery` class that handles most of the PhoneGap plugin boilerplate, and a `GetAction` class that interacts with the device's battery APIs.

   Here is the `Battery` class—you can see how similar it is to the Android implementation:

```
package com.phonegap.plugins;

import org.json.me.JSONArray;

import com.phonegap.api.Plugin;
import com.phonegap.api.PluginResult;

public class Battery extends Plugin {

    protected static final String ACTION_GET = "get";

    public PluginResult execute(String action, JSONArray args,
    String callbackId) {
        PluginResult result = null;

        action = (action == null) ? "" : action.toLowerCase();

        if (action.equals(ACTION_GET)) {
            result = GetAction.execute(args);
        }
        else {
            result = new
            PluginResult(PluginResult.Status.INVALIDACTION,
            "Battery: invalid action " + action);
        }

        return result;
    }
}
```

   All quite straightforward, I hope, if you followed along with the Android example.

5. Here is the `GetAction` class, which is more like the iOS plugin—a simple property access:

```
package com.phonegap.plugins;

import com.phonegap.api.PluginResult;

import org.json.me.JSONArray;
import org.json.me.JSONException;
import org.json.me.JSONObject;

import net.rim.device.api.system.DeviceInfo;

public class GetAction {

    private static final String
    PROPERTY_POWER = "power";

    public static PluginResult
    execute(JSONArray args) {
        PluginResult result   =
        null;
        String       property =
        getPropertyName(args);

        if (property.equals(PROPERTY_POWER)) {
            result = getPower();
        }
        else {
            result = new
            PluginResult(PluginResult.Status.INVALIDACTION,
            "Unknown property: " +
             property);
        }

        return result;
    }

    private static String
    getPropertyName(JSONArray
    args) {
        try {
            return
            args.getString(0)
           .toLowerCase();
        }
        catch (JSONException e) {
            return "";
        }
    }
```

```
    private static PluginResult
    getPower() {
        JSONObject json = new
        JSONObject();

        try {
            json.put("level",
            DeviceInfo.
            getBatteryLevel());
        }
        catch (JSONException e) {
        }

        return new PluginResult(PluginResult.Status.OK, json);
    }
}
```

There is a bit more boilerplate here for stylistic purposes—by using the Java JSON libraries, the objects are passed back to the callback as JavaScript objects, rather than as JSON strings to be parsed. But essentially, it's a very similar process to the iOS and Android roundtrip.

As of PhoneGap 1.0, your BlackBerry project will have a `plugins.xml` file in the www directory; as with Android, register the Battery plugin in that file:

```
<plugin name="Battery" value="com. phonegap.plugins.Battery" />
```

Load that up in your BlackBerry emulator and check out the results:

6. And as a nice bonus for all of your hard work, the BlackBerry emulator evens lets you futz around with the virtual device's battery power (unfortunately, the other platforms do not provide this), so you can enjoy the fruits of your work to your heart's content:

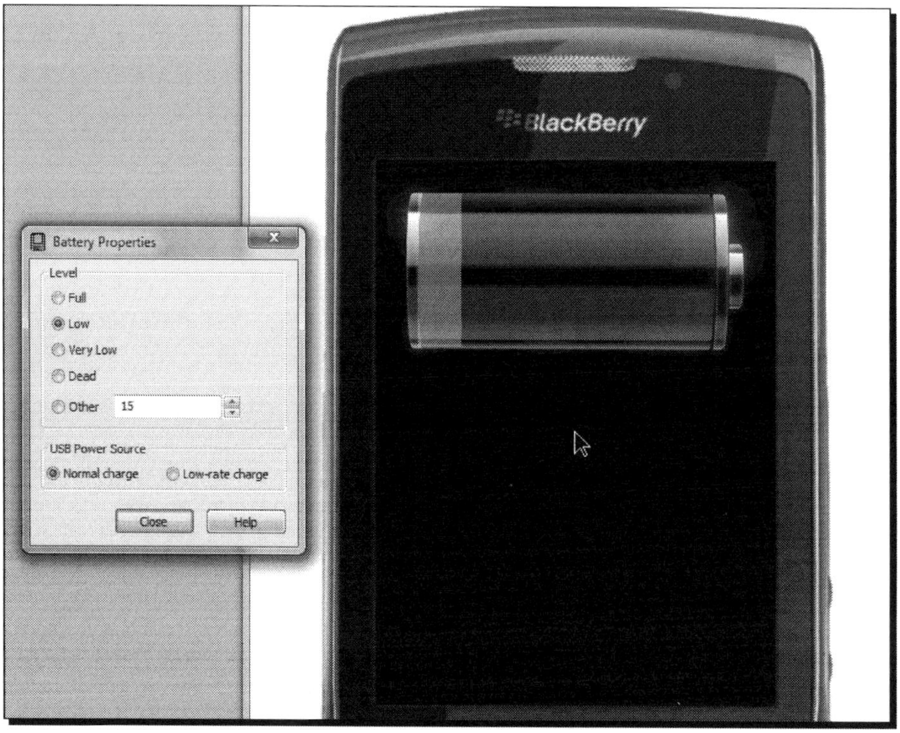

What just happened?

We took a bit of extra functionality that we had added to our application on iOS, and quickly ported it over to Android and BlackBerry, to extend PhoneGap with additional capabilities.

Although some of the intricacies of the PhoneGap plugin and PluginResult APIs take a little bit of getting used to, the net result is that you need to write very little JavaScript of native code to add functionality to your existing applications. The Android platform was an interesting example of a platform with a little bit of added complexity—there was a message based system for getting battery level, rather than a direct property access from the device—but this did not greatly impact the code that we had written.

PhoneGap plugins do not completely shield you from the intricacies of coding for a native platform—in the case of Android, some rudimentary knowledge of how Activities, Intents, Services, and BroadcastReceivers interact is very useful—but they do allow you to focus on finding quick solutions for important APIs, and integrate those quickly into your applications.

Do you need cross-platform plugins?

I've shown how to take a simple plugin and port it across three separate platforms, but there may be a valid concern over how worthwhile that exercise is. Typically, if a feature is something that would be useful for developers on all major platforms, it should be included in PhoneGap core.

One of the main selling points of the plugin infrastructure is to allow developers to quickly bridge native, single platforms features—iAds on iOS, cloud to device messaging on Android—into their regular PhoneGap applications, rather than waiting on the framework as a whole to integrate those capabilities. In most cases, if the feature does not work well cross-platform, it won't make it anywhere near standard PhoneGap. In the case of heavily demanded features, you may be able to find somebody on the mailing list or IRC who has had the same issue and worked towards fixing it, but there's no guarantee.

No limits

Once you start writing native code, you may realize that you can write anything you want. And you would be absolutely right—the plugin architecture of PhoneGap gives some light guidelines for developers, but any native code you want can be executed in the context of a PhoneGap application.

If your goal is to write cross-platform applications, I would be careful about which features you write as plugins and which ones you write using HTML5 and JavaScript. Web technologies, if all else fail, can run on the Web, and the Web runs just about everywhere. For simple device data, like our battery example, it's trivial to port between platforms, but any kind of involved plugin work, especially that involving user interface or view elements, quickly adds up when it has to be ported to multiple platforms. A good rule of thumb is to write something that works well using the tools available with web technologies, and optimize with native code if you encounter any bottlenecks in your approach.

Pop quiz: Writing PhoneGap plugins

1. Why should you start with a single JavaScript API for a plugin, and write your native implementations after that?
    a. Having a standard JavaScript API across all platforms increases code reuse
    b. Native code takes longer to write, and you don't want to get sidetracked
    c. If the native APIs change, you won't need to change your PhoneGap code

2. What is the main difference between plugins on Android and BlackBerry WebWorks?
    a. BlackBerry WebWorks requires all `PluginResult` objects to be JSON-based
    b. BlackBerry runs on stock Java, while Android is based on the Dalvik VM
    c. The underlying platform APIs are very different
3. Can you write native code for your application without using plugins?
    a. No: A PhoneGap application can only access the native side through the plugin interface
    b. Yes: Just use the same bridging techniques as the main PhoneGap APIs are implemented with

Summary

We managed to have a whirlwind tour around PhoneGap plugins, and taking our PhoneGap applications outside of the PhoneGap APIs. In particular, we learned to:

- Integrate existing plugins into our application, to extend the functionality presented by the framework
- Write a simple plugin for iOS, sending data back to our webview
- Port that plugin to the Java-based PhoneGap platforms, for cross-platform battery action

We're winding down our trip around the PhoneGap, uh, software; it's time to look at one of the major challenges facing developers of mobile applications. How do you effectively sync your application, so it can receive new data from the web, and still work smoothly offline?

11

Working Offline: Sync and Caching

Up to this point, we've mostly looked at PhoneGap applications in isolation, as relatively simple and self-contained units of functionality, interacting with native APIs. While this is certainly a feasible approach, mobile applications gain a great deal in expressiveness and capability when combined with remote web services. Whether it's an existing service (such as the Twitter API, which we've touched on in passing), or one developed alongside your application, tying your work on mobile devices to a remote service allows for vast possibilities in application development.

In this final chapter, we will look at some of the approaches for working with a remote web service. In particular, we will see how to:

- Write a simple service to tie with a mobile application
- Cache updates from the remote server, and sync those with new updates to keep users up to date

We're going to start off by writing a tiny little web service to work with our application—a citizen news site called, imaginatively, **New Stories**.

Ruby and Sinatra

Many front-end web developers may not have implemented a web service before, leaving that boring stuff to the neckbeards and the sysadmins. The truth is, however, that it's an incredibly simple process, especially with the tools available to a modern web developer.

There are many different tools available for writing web services, but our example will use the **Sinatra** web framework, written in the **Ruby** programming language. Sinatra allows for clean, expressive definitions of responses to web requests, and allows us to write our proof of concept web service code with a minimum of boilerplate. It's often described as domain-specific language for writing web services—closer to writing declarations with imperative programming. I highly recommend it for any of your lightweight web processing needs.

If you're running on Mac OS X or Linux, you'll already have Ruby installed; if you're running Windows, visit `http://www.ruby-lang.org` to get the Ruby installer, and the **RubyGems** package manager.

To install Sinatra, and the JSON library we will use in our example, run the following command:

```
gem install sinatra json
```

If you're on Windows, the regular command prompt should work fine, rather than using Cygwin.

For more information on Sinatra, you can find extensive documentation and details at `http://www.sinatrarb.com`.

Time for action – A news site, with an API

1. Let's get something started as quickly as possible with Sinatra—create a new file called `server.rb`, and enter the following:

```
require 'rubygems'
require 'sinatra'

get '/' do
    "Hello World"
end
```

The first two lines ensure we have access to Ruby's package manager, RubyGems, and the Sinatra library. The rest of file says that, when our server receives a GET request for the path /, we should respond with the string **Hello World**.

2. Run this file by the command `ruby server.rb`. You should see the following output:

```
== Sinatra/1.2.6 has taken the stage on 4567 for development with backup from Thin
 >> Thin web server (v1.2.7 codename No Hup)
 >> Maximum connections set to 1024
 >> Listening on 0.0.0.0:4567, CTRL+C to stop
```

Navigate your web browser to the URL at the bottom to see your very first web service in action:

3. There's not much there now, but let's give it some time. We're going to update our server to generate some dynamic views, rather than always returning the string **Hello World**. What we want is to display a list of news stories.

Here's how we'll change the `server.rb` file:

```
require 'rubygems'
require 'sinatra'
require 'erb'

@@stories = [
  {
    :time => Time.now.to_i,
    :title => "Outbreak in Europe",
    :body => "There has been a terrifying outbreak in Europe. Be
    forewarned!"
  }
]

get '/' do
  erb :index, :locals => { :stories => @@stories }
end
```

At the top, we're calling `require` for another library, the templating library **erb**, or embedded Ruby. Next, we're defining a class variable (with that curious **@@** syntax) to hold all of our news stories, with a sample story to flesh it out. We've then changed the handler for the / request to render, with erb, the index template (which we have to write), passing the `@@stories` class variable as a local variable.

One thing to note is that, for the sake of brevity and simplicity, all of our stories are being saved in an array in memory. Were this a serious application, those stories would of course be backed up to a persistent database.

4. Now for that index template. Create a directory called `views`, and then a file in there called `index.erb`, and fill it with the following:

```
<h1>News Stories</h1>

<% stories.each do |story| %>
  <h2><%= story[:title] %></h2>
  <p><%= story[:body] %></p>
  <p><small>Reported at <%= story[:time] %></small></p>
<% end %>

<h3 style="margin-top: 50px">Submit your own story:</h3>
```

This should look similar to the Mustache templates we've been writing throughout the last few chapters; we're iterating through the list of stories and printing each one out as HTML. Restart the server (typically, press *CTRL-C* to end the process and then the up arrow to call the last command again), and reload your browser.

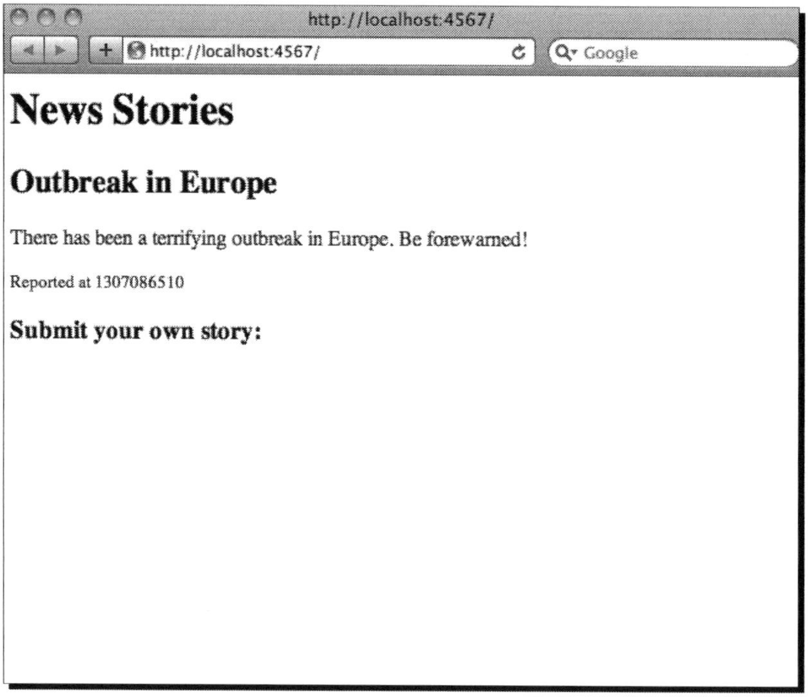

It's pretty ugly, sure, but it gets the job done.

5. We next need a mechanism for posting new stories to the site—sure, an outbreak in Europe is big news, but do we have anything else? Let's write the front-end code first, and add the following to `views/index.erb`:

```
<form action="/story" method="post">
  <p><b>Title</b>
    <input type="text" name="title" /></p>

  <p><b>Body</b>
    <textarea name="body"></textarea></p>

  <input type="submit">
</form>
```

We can see from the `action` and `method` attributes of the form that we're posting to the path `/story`; let's add the appropriate Sinatra code in `server.rb` to handle that:

```
post '/story' do
  new_story = {
    :time => Time.now.to_i,
    :title => params[:title],
    :body => params[:body]
  }

  @@stories.unshift new_story

  redirect '/'
end
```

If it's not clear, this method reads the parameters from the form, builds a suitable story object, and adds it to the list of stories. It then redirects the user back to the home page, which will be updated with the new story.

Caveat lector: this is very bad code for anything approaching the public Internet. Don't take user input and put it directly in your page's markup! Since you will be the only person accessing this web service, it's not a big deal, but this code should never be used on a production website.

Restart the server, and try entering a few stories from your browser to test out the service:

So far so good—we've buried the depressing **Outbreak in Europe** story with a far more exciting **Brand New Console** one.

6. For the final point, we want to be able to access the news stories in a format that we can easily process. For best results, we'll take JSON.

   To get our server serving up JSON, we'll need to first of all require the JSON library, by adding this line to the top of our `server.rb`:

   ```
   require 'json'
   ```

7. Then, add a route handler to present our stories as a JSON formatted list. To keep things easy to handle, we'll limit this to the first five stories:

   ```
   get '/json' do
     content_type :json
     @@stories[0,5].to_json
   end
   ```

Note how easy it is to set the `content-type` header using Sinatra—it looks more like a declaration than a function call. Now, navigate to `http://localhost:4567/json` in your browser to confirm that everything looks okay.

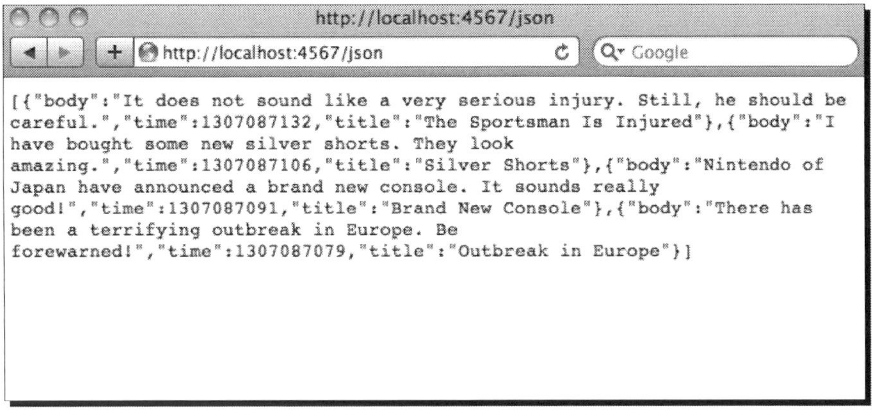

8. Now for our mobile application. If you've followed the preceding chapters, this should be simple enough—create a new www directory, and copy in `xui.js` and `mustache.js`. Create an `index.html` and add the following:

```
<html>
    <head>
        <title>News From The Internet</title>
        <link rel="stylesheet" href="style.css" />
        <meta name="viewport" content="width=device-width,initial-
         scale=1.0" />
    </head>
    <body>
        <h1>News From The Internet</h1>
        <div id="news-container">
        </div>
    </body>
    <script src="phonegap.js"></script>
    <script src="xui.js"></script>
    <script src="mustache.js"></script>
    <script src="app.js"></script>
</html>
```

Now create a `style.css` file, and add the styles:

```
body {
    background: #56F;
    color: white;
    font-family: Helvetica Neue;
}
h1 {
    color: #FF5;
    text-align: center;
}
p {
    color: black;
}
```

Finally, `app.js`:

```
var newsStory = "<h2>{{ title }}</h2><p>{{ body }}</p>";

x$(document).on('DOMContentLoaded', function () {
    x$().xhr("http://localhost:4567/json", function () {
        var storiesArray = JSON.parse(this.responseText),
            storiesMarkup = "",
            storyHtml;

        storiesArray.forEach(function (story) {
            storyHtml = Mustache.to_html(newsStory, story);
            storiesMarkup += storyHtml;
        });

        x$('#news-container').inner(storiesMarkup);
    });
});
```

9. Everything should be clear: we wait for the DOM to be ready, query our web service for the latest stories, and then render those to the DOM, using Mustache for templating.

Check things in Safari—you're now a full-stack web developer!

What just happened?

We were able to use the Ruby programming language and the Sinatra web framework to write a simple web service, that enables applications to query a list of timestamped news stories and display them in a mobile-optimized format, in only a couple of dozen lines of codes. We were then able to use our existing client-side skills to connect to this web service, and display the data appropriately.

With a little more effort, we could easily modify our application to handle requests for a single news story, for a paginated set of news stories, or to delete news stories from the server. But for now, this simple service suits our purposes.

Alternatives to Sinatra

Sinatra has been called the most cloned of all open source projects; its syntax of defining behavior through HTTP methods and route paths has proved immensely popular. We're great fans of using it for this kind of purpose, although you may want a more comprehensive web framework if your goals are suitably complex.

If you want to work with JavaScript on the whole stack, and use `node.js` for writing your server, the **Express** framework offers a Sintra-style syntax in JavaScript. For PHP developers, the **Limonade** framework has proven to be a popular option for lightweight web services, and if you're using Python, you can try `web.py`. Most other languages or runtimes will offer similar solutions.

Lightweight web services like this one may seem useful for tutorials, but the purpose of writing them in your day-to-day work may be a bit more elusive. If you're a front-end web developer working on a large project, you typically won't have a great deal of control over what technology stack the server uses, or how the individual requests are formatted and processed.

In this scenario though, it's often useful to whip together a small web service for your development needs, especially if the stability and availability of the remote service is in question. Once you have an idea of the format of the responses you'll be dependent on, you can use a lightweight framework like Sinatra to stub out these responses, allowing you to continue working whether the external service, or your own environment, is offline at that time. It's an extremely powerful option to have in your toolchain.

Pop quiz: A simple web service

1. Why does the service offer the news stories in JSON format, as well as HTML?
    a. JSON is easier for applications to parse and process than HTML.
    b. JSON representations are typically smaller than their HTML counterparts, so less bandwidth is required to transfer them.
    c. The HTML may contain content that is unnecessary for a consuming application, such as the HTML form in our example.
2. Is our **News Stories** web service ready for a production deployment?
    a. Yes: it serves its purpose of displaying news stories and accepting new ones.
    b. No: it has a number of glaring security flaws, as well as no persistent storage for stories.

3. What is the best approach for loading our news stories from our service?
    a. Using an `XmlHttpRequest`, as in the example above.
    b. Using `script` tags, also known as JSONP.
    c. Rendering the service's HTML files directly in our application.

Caching new stories

Way back in *Chapter 3, Mobile Web to Mobile Applications* we briefly covered the two feasible options for offline storage on mobile devices: **localStorage** and **Web SQL**. For lightweight data storage, things like user preferences or application state, localStorage is more than sufficient. However, when dealing with remote web service, you often have a more complex data model to deal with. In this case, it's a good idea to take a look at Web SQL.

Note that WebSQL is not supported on Firefox—make sure you're using a WebKit-based browser when testing this code.

Once we set up a small database for our mobile news client, we can see how easy it is to cache data from the web service, enabling our application to successfully run offline.

Time for action – Caching stories in a local database

Since the interface for using Web SQL is quite verbose, we're going to create a separate JavaScript file to hold all of our database specific code.

1. Create a file called `database.js` and add a reference to it in your `index.html`:

```
<script src="phonegap.js"></script>
    <script src="xui.js"></script>
    <script src="mustache.js"></script>
    <script src="database.js"></script>
    <script src="app.js"></script>
```

The first thing we need to do when using Web SQL is get a reference to our database, using the `openDatabase` function:

```
// open the database
function initializeDB() {
    var localDatabase = openDatabase(
            "internetNews",              // short name
            "1.0",                       // version
            "News From The Internet",    // long name
            5000000                      // maximum size in bytes
        );

    return localDatabase;
}
```

You can call `window.openDatabase` on most platforms to emphasize that you're calling a global function; unfortunately, this doesn't behave predictably in all mobile environments. Calling `openDatabase` as `openDatabase` does, even if it's a little gross.

Most of the Web SQL API is quite unintuitive—lots of ordered and required parameters. I've commented the ones we're using—a short name, a verison , a long name, and a maximum size for the database. I've specified five megabytes (or thereabouts) for our database's maximum size—this is the most we can reliably get in most environments.

Can you believe this API is deprecated (refer back to *Chapter 3, Mobile Web to Mobile Applications* for more details on this)?

2. To test that our function works correctly, open `index.html` in Safari and run the `initializeDB()` function from Web Inspector. If you switch to the **Storage** tab in Web Inspector, you should see your newly created database available:

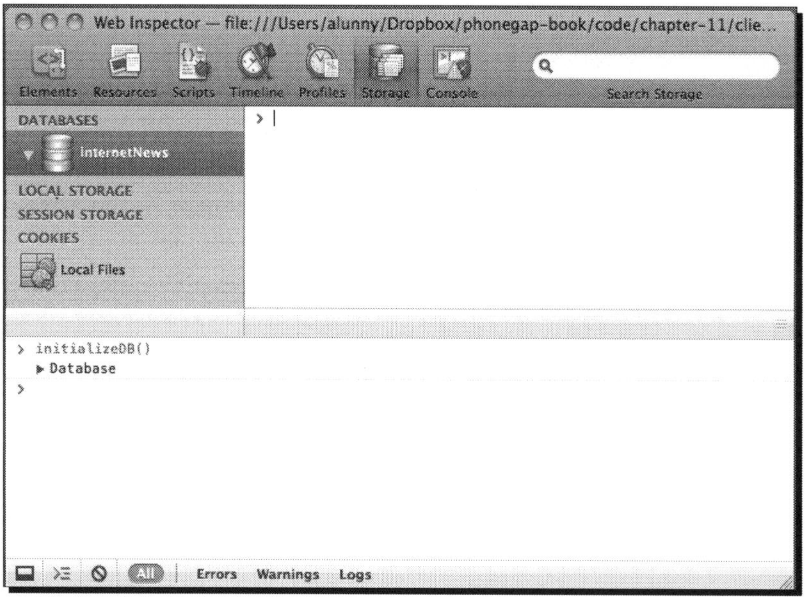

3. Anyone familiar with SQL will know that our next step is to create a table to store our records. To do this, we'll need to start a transaction on our `localDatabase`, and execute the SQL query asynchronously. Here's how this looks:

```
function createStoryTable(db) {
    var query = "CREATE TABLE IF NOT EXISTS stories " +
        "(title NVARCHAR(25), time DATETIME PRIMARY KEY, body
        TEXT);"
```

```
    db.transaction(function (trxn) {
        trxn.executeSql(
            query,  // the query to execute
            [],     // parameters for the query
            function (transaction, resultSet) { // success
            callback
                console.log('success');
            },
            function (transaction, error) { //error callback
                console.log(error);
            }
        );
    });
}
```

We have to write the query (`CREATE TABLE...`) manually, knowing before hand the data types for each of the columns in our table. There's no programmatic way to get most of this information (for example, which data types are supported by the database); you'll have to look at the documentation for the particular release of SQLite that is embedded in the particular web view you're working with. We're setting the `time` field to be our primary key, for simplicity's sake; in a production application, you may well have an `id` field to uniquely identify records in the table.

The `database.transaction` function works asynchronously, taking a callback function that is passed a transaction object which you can execute SQL on asynchronously. I've omitted passing an error callback to this initial transaction call, in the interest of maintaining some semblance of sanity.

The `executeSql` function takes four parameters: a query, optionally parameterized; the parameters for that query, if it is parameterized; and two callbacks, for success or error. Each of the final two callbacks is also passed the transaction as an initial parameter, allowing for an unending stream of callbacks to spread across the screen with every increasing indentation. We're not interested right now in hooking into these callbacks—we just want to fire and forget, essentially.

Reload this page in Safari and try running the following code from Web Inspector (you need to call `initializeDB` again to grab a reference to the database):

```
myDB = initializeDB()
createStoryTable(myDB)
```

Looking in the Storage pane once more, you should see the table appear:

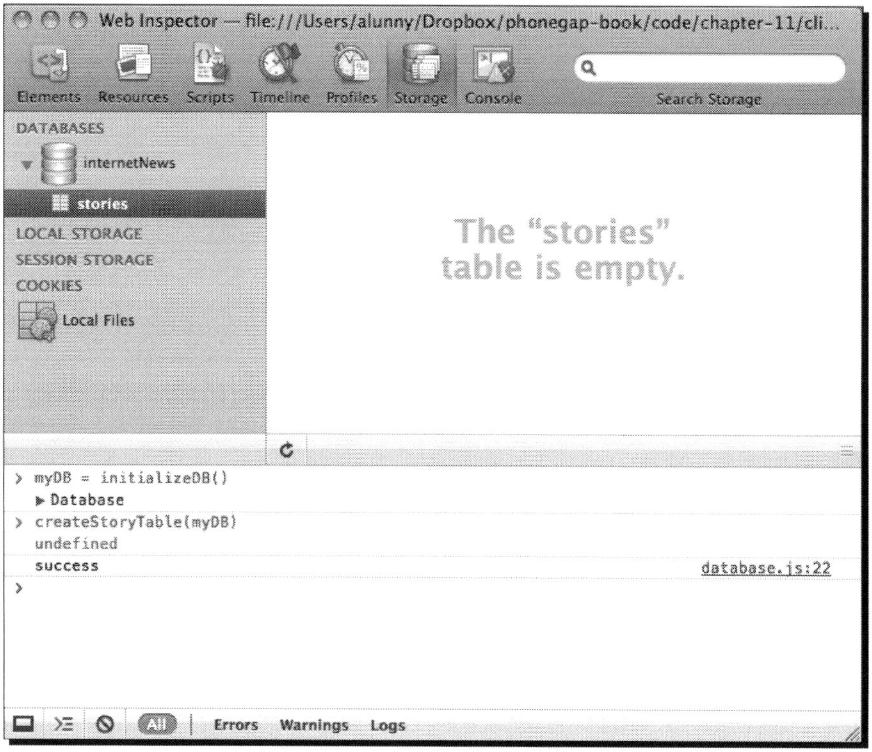

4. The next step is to make some inserts into the table. We're going to write a simple function that takes the JSON-parsed story objects we receive from the server, and insert those into our `stories` table:

```
function insertNewStory(db, story) {
    var query = "INSERT OR REPLACE INTO stories (title, time,
    body) " +
        "VALUES (?,?,?);"

    db.transaction(function (trxn) {
        trxn.executeSql(
            query,
            [story.title, story.time, story.body],
            function (transaction, resultSet) {
                console.log('success');
            },
            function (transaction, error) {
                console.log(error);
            }
```

```
        );
    });
}
```

The format is very similar to the `transaction`/`executeSql` calls for our `createStoryTable` function. We pass in a reference to our database object, along with the story object we have received from the server, and then build a query for inserting it into our database.

5. To see the writing in action, we're going to modify our `app.js` file to create the table, if it's not present, and then insert all of the stories we get from the server. Here's our modified `DOMContentLoaded` handler:

```
x$(document).on('DOMContentLoaded', function () {
    var myDB = initializeDB();

    createStoryTable(myDB, function () {
        x$().xhr("http://localhost:4567/json", function () {
            var storiesArray = JSON.parse(this.responseText),
                storiesMarkup = "",
                storyHtml;

            storiesArray.forEach(function (story) {
                insertNewStory(myDB, story);
                storyHtml = Mustache.to_html(newsStory, story);
                storiesMarkup += storyHtml;
            });

            x$('#news-container').inner(storiesMarkup);
        });
    });
});
```

Note that we're now passing a callback function to `createStoryTable`, so we'll need to modify that function in a couple of places:

```
function createStoryTable(db, callback) {
    var query = "CREATE TABLE IF NOT EXISTS stories " +
        "(title NVARCHAR(25), time DATETIME PRIMARY KEY, body
        TEXT);"

    db.transaction(function (trxn) {
        trxn.executeSql(
            query,  // the query to execute
            [],     // parameters for the query
            callback,
```

```
                function (transaction, error) { //error callback
                    console.log(error);
                }
            );
        });
    }
```

6. Reload the page in Safari, and you should be able to see the table automatically populated in the Web Inspector:

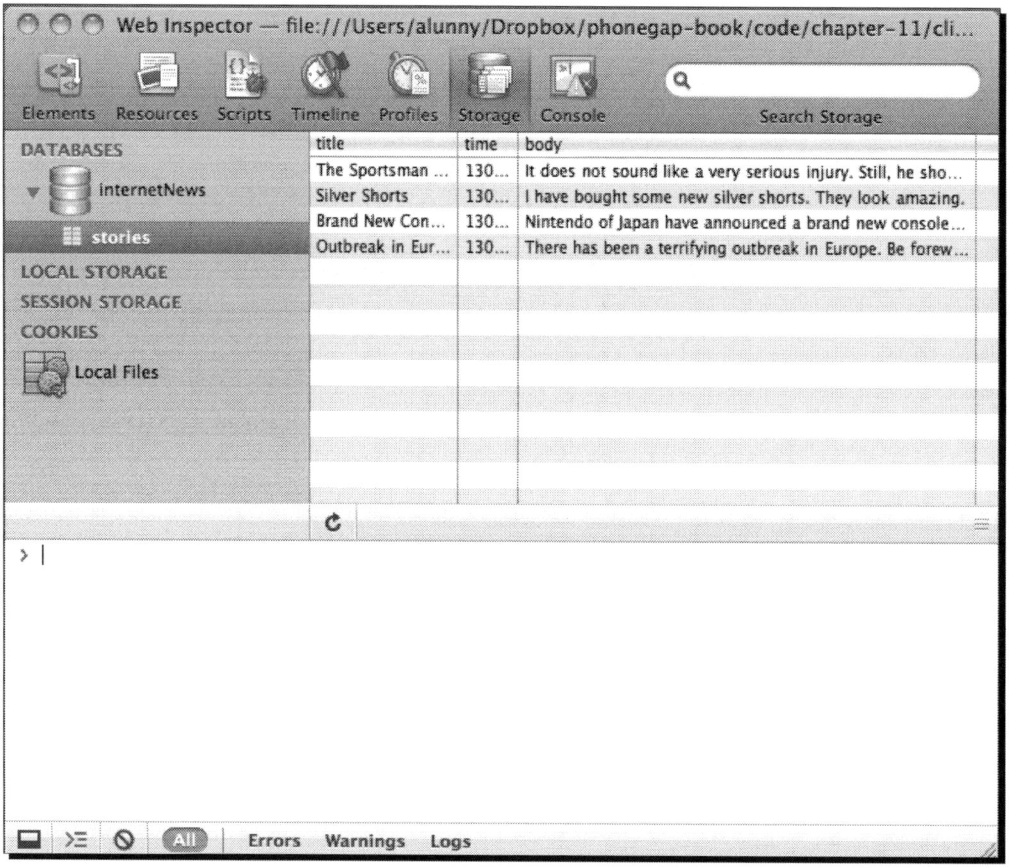

7. The final thing we need to do is ensure that rendering works just as well with or without online access. If you're still running our Sinatra service in a terminal, you should kill the process now, and reload Safari. Everything looks a bit blank, even though the records are still present in our local database:

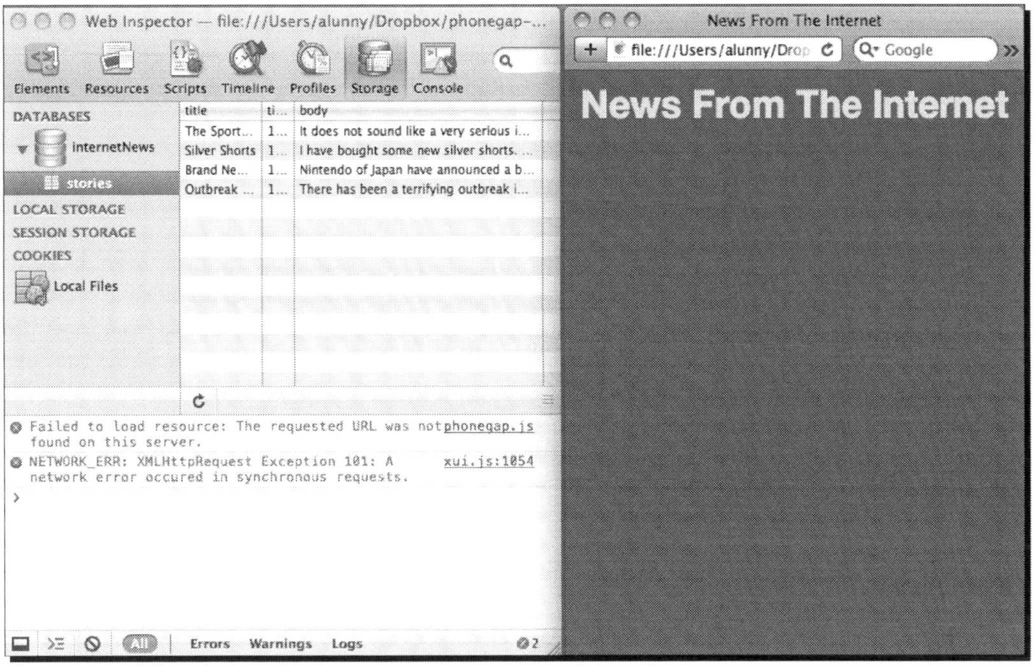

Obviously this is unfortunate; in this case, we need to retrieve the records from the local database and render those directly. Firstly, let's write a function that gets all of the stories out of the local database, and returns them in the array format we're familiar with:

```
function getLastStories(db, callback) {
    var query = "SELECT * FROM stories LIMIT 5;"

    db.transaction(function (trxn) {
        trxn.executeSql(
            query,  // the query to execute
            [],     // parameters for the query
            function (transaction, resultSet) {
                var i = 0,
                    currentRow,
                    stories = [];

                for (i; i < resultSet.rows.length; i++) {
                    currentRow = resultSet.rows.item(i);
                    stories.push(currentRow);
                }
```

```
                callback(stories);
            },
            function (transaction, error) { //error callback
                console.log(error);
            }
        );
    });
}
```

This is similar to the last two SQL querying functions that we've written—the major change is in the success callback function. This time, we actually are interested in the contents of the `resultSet` parameter, which is a little bit of a strange object. Unlike most JavaScript objects, the members are accessed in iterator-style—we use the `item` function to pull out each row individually. We then pass an array of these rows back to our callback function, allowing these to be rendered in the same way content from the server is.

8. Finally, let's refactor our `app.js` code a little more to accommodate this use case. We're going to pull the rendering code out into a separate function, and then call either the online or offline code to get the list of stories, depending on whether we get back a useful response from the server:

```
x$(document).on('DOMContentLoaded', function () {
    var myDB = initializeDB();

    function renderStories(stories) {
        var storiesMarkup = "",
            storyHtml;

        stories.forEach(function (story) {
            storyHtml = Mustache.to_html(newsStory, story);
            storiesMarkup += storyHtml;
        });

        x$('#news-container').inner(storiesMarkup);
    }

    createStoryTable(myDB, function () {
        x$().xhr("http://localhost:4567/json", {
                async: true,
                callback: function () {
                    try {
                        var storiesArray =
                        JSON.parse(this.responseText);
```

```
                        storiesArray.forEach(function (story) {
                            insertNewStory(myDB, story);
                        });

                        renderStories(storiesArray)
                    } catch (e) {
                        // failed to retrieve data
                        getLastStories(myDB, function (stories) {
                            renderStories(stories);
                        })
                    }
                }
            });
        });
    });
```

Note that we did have to modify our `x$.xhr` call a fair bit—the behavior is a little different depending on what kind of external connection you're expecting.

Reload in Safari and you should see our news stories right back on the screen, where they belong.

What just happened?

Frankly, not an awful lot. WebSQL's incredibly verbose interface made getting even basic caching in place quite an ordeal.

There's also much that we did not cover, even for the simple caching workflow. How should your application handle schema migrations on the remote server, if there are extra fields to take care of? Should you seed your application with initial data, or download everything at runtime? How should you manage compacting the database, to remove old data? These are all interesting problems that aren't amenable to a one-size-fits-all solution—you will need to assess what the best approaches are for your particular application, and how much additional complexity you're willing to put up with, on the server and on the client.

Managing application initialization

One important consideration when building this kind of application is how much work should be done at initial runtime. Even with our sample application, we're beginning to stretch the bounds of how much logic can be performed on application initialization.

Roughly speaking, there are three events that we're concerned about on application initialization:

- **Script loading**: This isn't an event you can subscribe to, but I'm referring to the time when the web view parses and executes your JavaScript files. Throughout the book, I've mostly used this event to define functions and subscribe to other events. However, you may wish to experiment with calling more of your application logic at this time.
- **DOMContenLoaded**: This means the `document` has been rendered, and you can start modifying the contents with some confidence. Typically, any rendering you have to do should have to wait until this point, for safety's sake.
- **deviceready**: As we've seen in previous chapters, this means that PhoneGap specific APIs are now available. For example, if we wanted to use `navigator.network.isReachable` to see if we could reach our remote server, rather than just putting an error handler on our callback function, we would have to wait for `deviceready`.

Once you integrate your application closely with a remote service, it's worth doing some benchmarking to see exactly how much work you should be doing at each of these stages. In a complex sync scenario, you could be doing any or all of the following:

- Check for a remote service's availability
- Query the remote service for any new updates
- Save new updates to a local database
- Save new media files to the local filesystem, using PhoneGap's File API or a custom plugin
- Read updated data from the local database
- Read updated media from the local filesystem
- Update the remote service with the state of the mobile application

And so forth. Add to this the fact that the SQLite interface (and also the forthcoming indexedDB interface) is controlled chiefly by asynchronous function calls, as will be the XHRs to remote servers and the PhoneGap calls you're making, and you could have quite a complex mess to deal with.

The best strategy is good old-fashioned diligence: keep an eye on how your application performs in every different scenario, and focus on providing responsiveness for the user at all times.

Have a go hero

If the preceding examples were not complex enough for you, it might be worth looking at extending our little server to allow for writes from our mobile application also. How do you handle potential conflicts? Can you ensure the timestamp matches on both the server and the client? When should your application update the remote server, and how should it know whether the server has received a given update or not?

Summary

We made a good first pass at integrating our mobile application with a remote service. We were able to:

- Write a simple web service, and populate our mobile application entirely from that content
- Cache the content to a local database, and then render the same content without a network connection available

That concludes our chapter on offline strategies, and our initial trip through PhoneGap. Hopefully you've enjoyed reading about PhoneGap, and are itching to get started writing some PhoneGap applications of your own. Good luck!

A

Deploying to iOS

For various political reasons, the iOS platform requires a bit of effort to get our application from a running simulator onto an actual device. It's the sort of busywork that is neither exciting nor valuable, but it can be a major sticking point for a lot of developers.

Note: deploying to a device requires a $99 Developer Certificate from Apple. If you wish to sell any applications, you will need to complete this as well.

Time for action–deploying to a device

1. First, you'll need to generate a Certificate Signing Request. Open Keychain Access (in your `/Applications/Utilities` directory) and start the Certificate Request wizard (see the following screenshot). Your developer name will be based on the one you registered your account with.

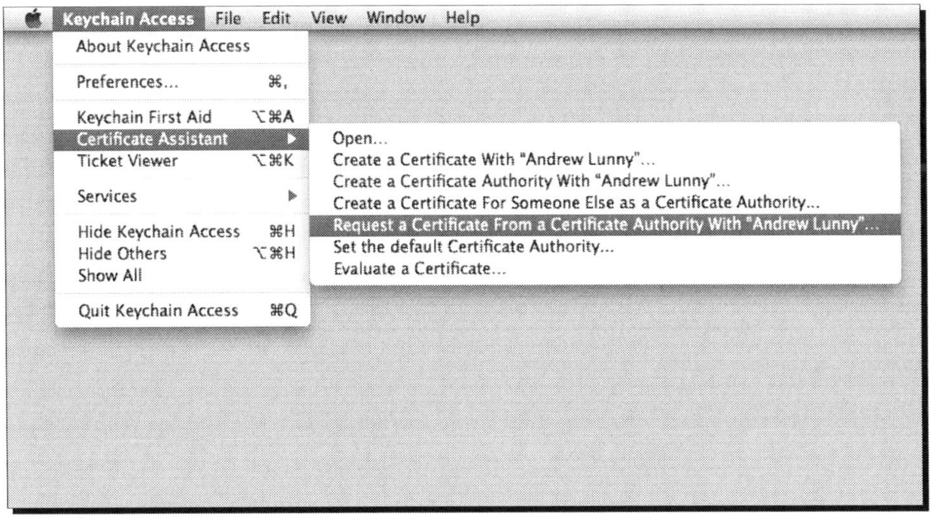

2. Fill out the e-mail you used to register for the Developer Program, and save the **Certificate Signing Request** (**CSR**) to disk. If prompted, set the key size to `2048` bits and the algorithm to RSA.

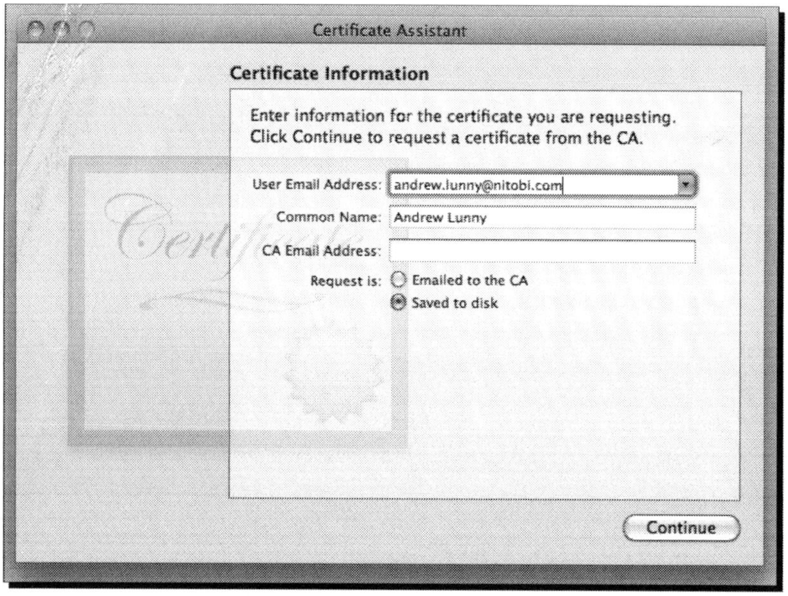

3. Log in to the **iOS Provisioning Portal** at `https://developer.apple.com/ios/my/overview/index.action` to upload your CSR, and then download the signed certificate once it appears (you may need to reload the page).

4. Once the certificate is downloaded, double-click it to install. You should see the certificate present in Keychain Access as shown in the following screenshot:

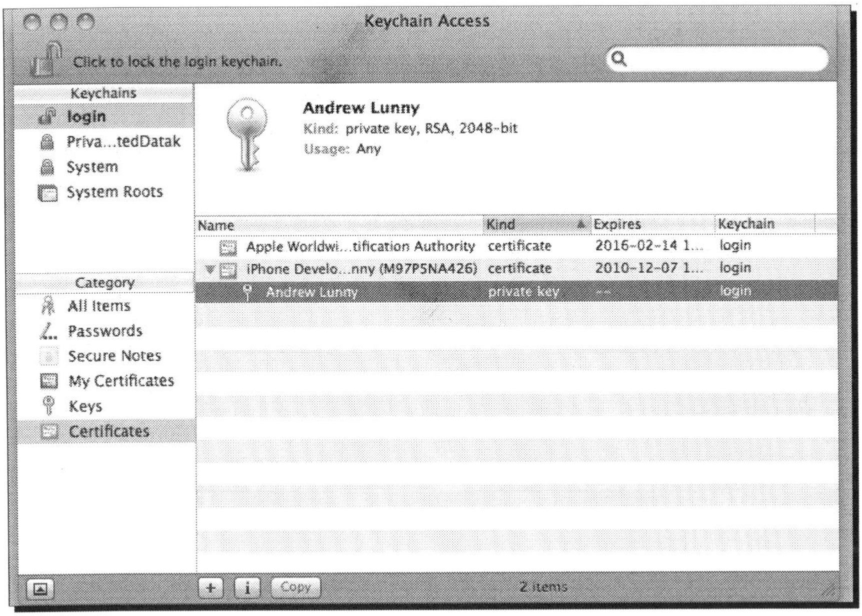

5. The next step is to register a device with the iOS Provisioning Profile–you'll need an iPhone, iPod Touch, or iPad here. Plug your device into your Mac's USB port, and open the **Organizer** in Xcode. You should see your device name with a green light next to it–in my case, I've connected my **palo alto pod iPod touch**. Your own device will have a name that you have specified.

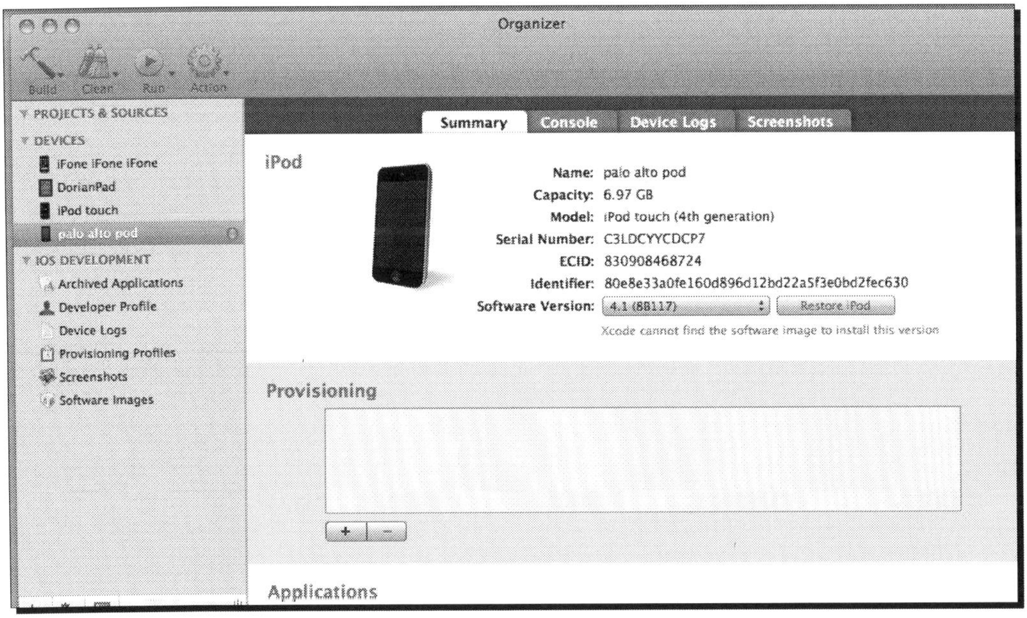

6. Copy the **Identifier**, then head back to the iOS Provisioning Portal, and hit the **Add Devices** button on the top right. Enter an identifiable name and the device ID.

7. The next step is to set up some App IDs, from the conveniently named **App IDs** link on the left-hand side of the iOS Portal.

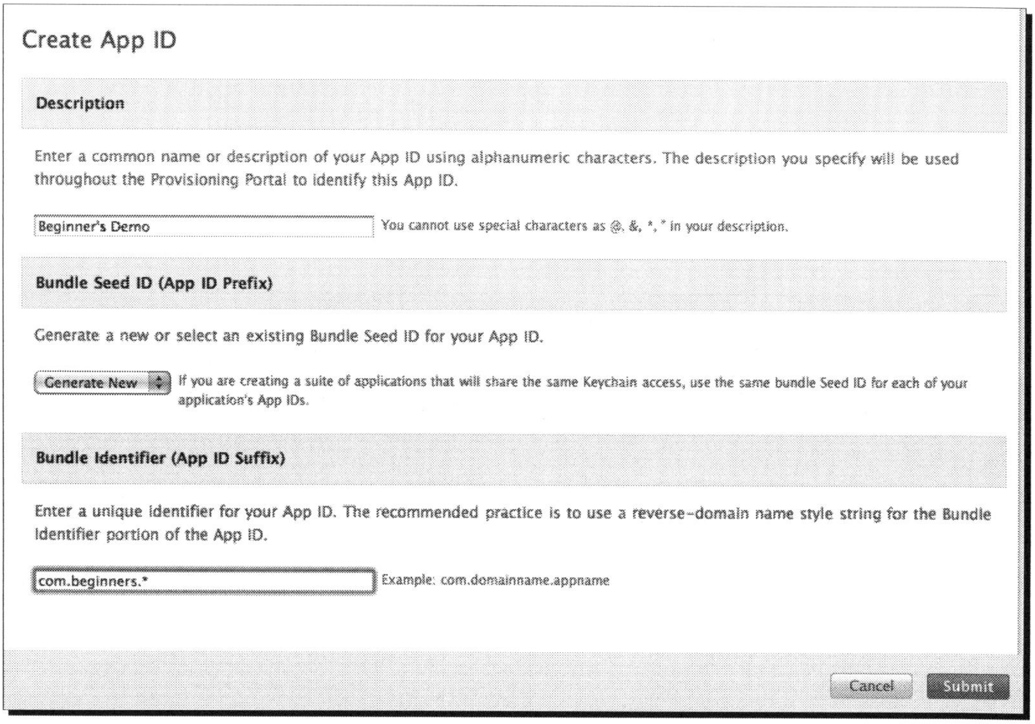

8. The form is pretty straightforward, but we want to make sure we can install multiple applications with the same provisioning profile, so be sure to use a wildcard name for the bundle identifier field. This will allow us to install, say, `com.beginners.appone`, `com.beginners.apptwo`, and so on. These identifiers are not related to Java packages, nor are they used for importing code—they're just for identifying your applications.

9. The final step is to tie the Developer Certificate, Device ID, and App ID together with a provisioning profile. Again, this is a simple form—choose a unique name, select the certificates you wish to use, the application identifiers, and all devices, as shown in the following screenshot:

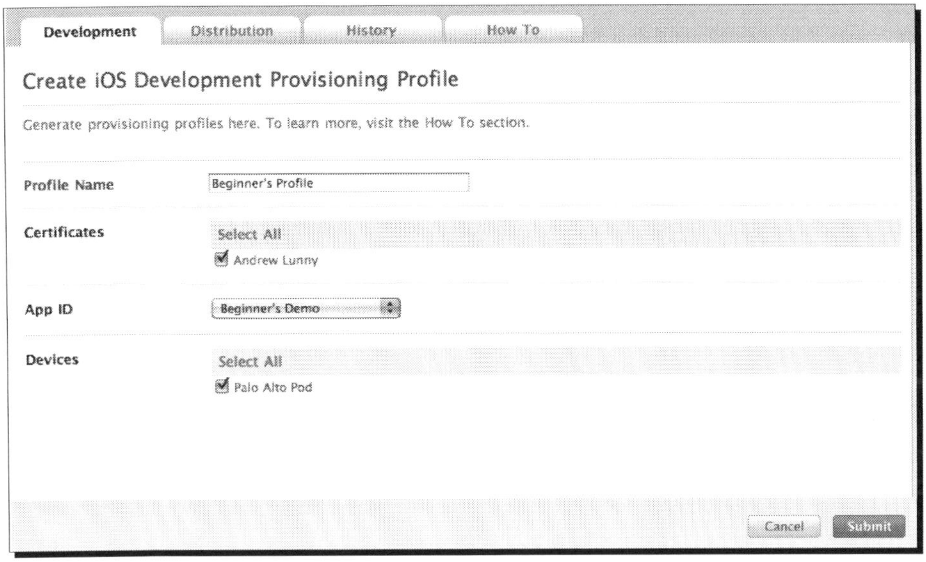

10. Download the `Beginners_Profile.mobileprovision` file and double-click on it. As long as your device is still connected, the profile will be installed on it as expected.

11. Go back to our `FirstApp` Xcode project (sure seems like a long time since we've been there). We'll need to edit the **FirstApp-Info.plist** file so that the App ID matches the one in the provisioning profile:

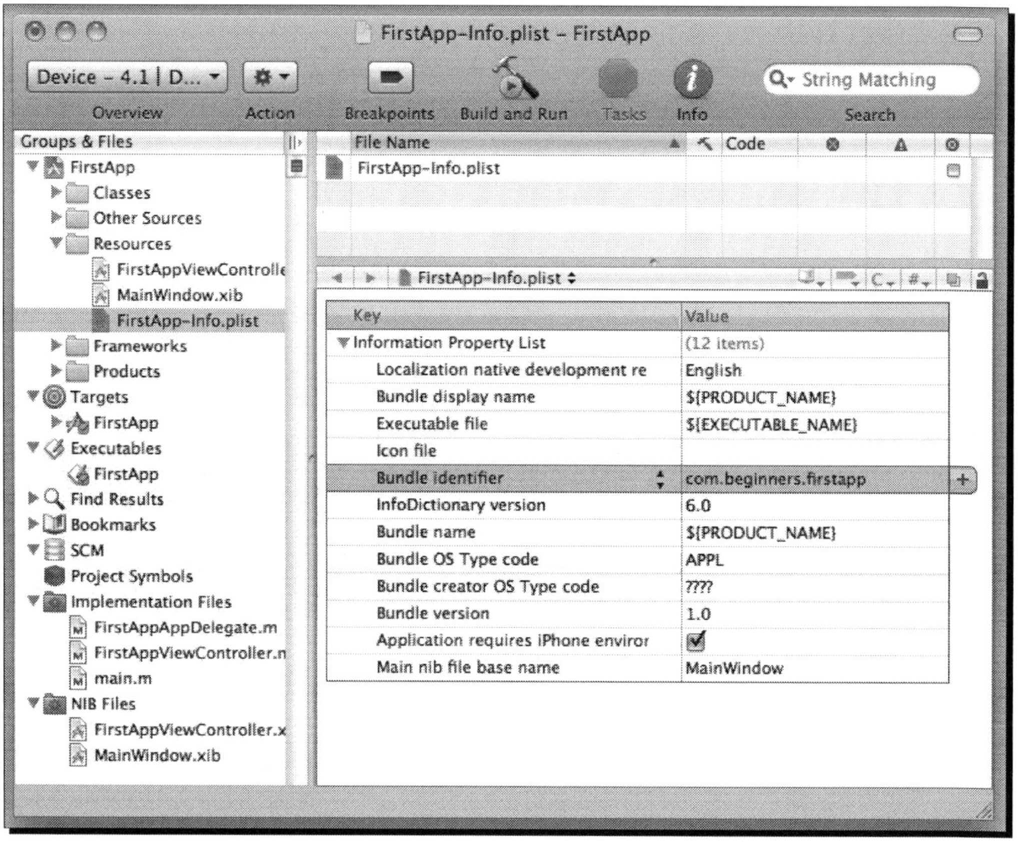

12. Set your **Code Signing Identity** for the application; double-click on the blue **FirstApp Xcode** project icon on the top left, and then change the **Code Signing Identity** field. If all has gone well, your name and ID should be in the pop-up list.

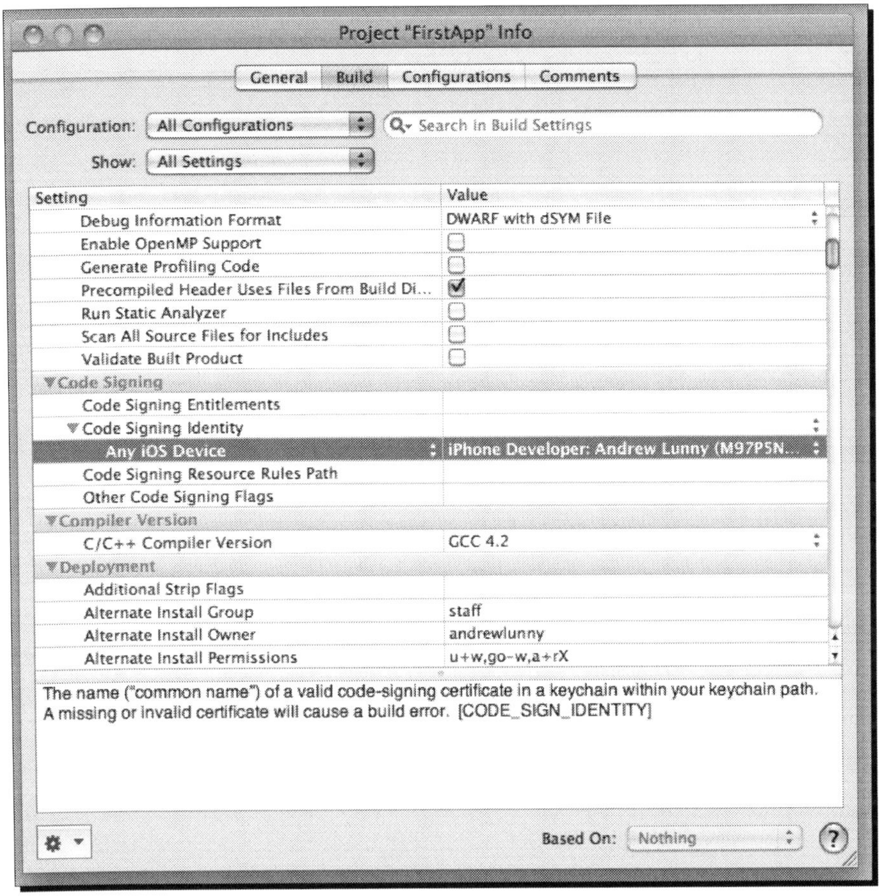

13. Ensure the target in the top left is set to **Device**, and the correct device is selected:

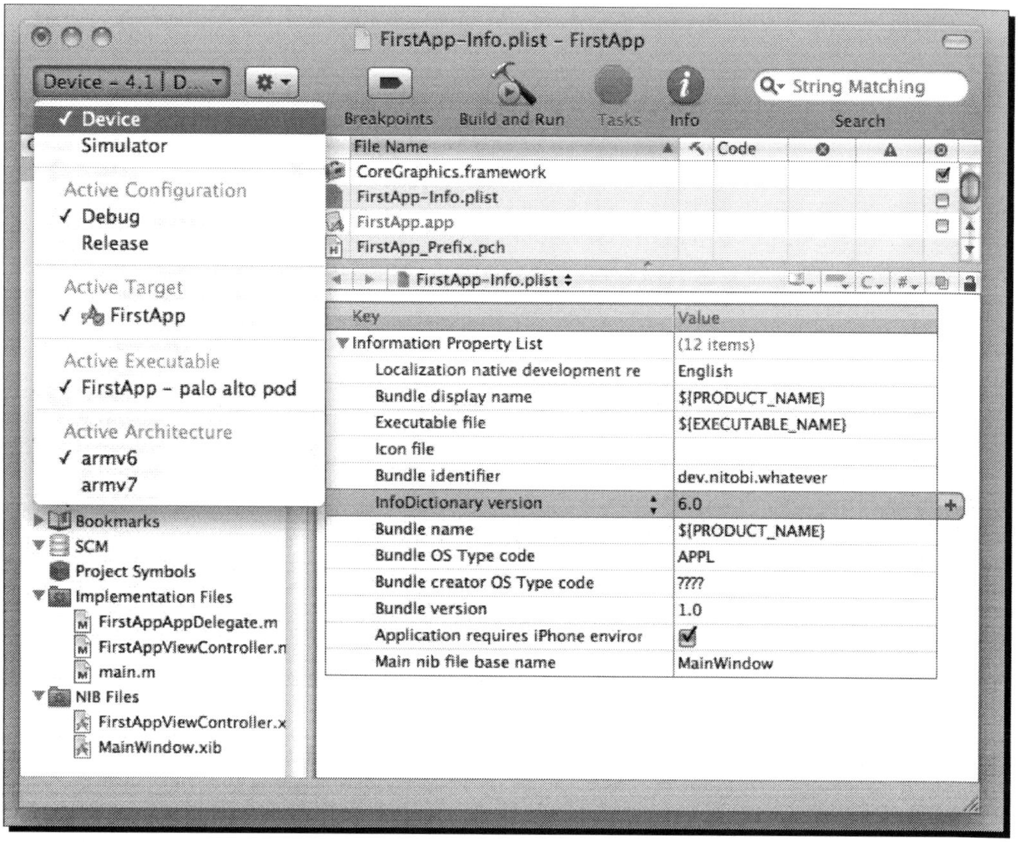

14. Bask in the glow of your beautiful grey screen application.

What just happened?

Well for one thing, we became a lot more familiar with the pain and bureaucracy that is iOS development. Whatever you think of the iOS development cycle (and personally, I'm a big fan of the SDK itself), there's an awful lot of ceremony involved in getting a viable setup up and running.

Thankfully, these steps do not need to be repeated often— that is, they only need to be done once for each Mac you wish to use for development. Here are some good guidelines for maintaining your sanity:

- Use wildcard App IDs for your provisioning profile—some developers even go as far as using the simple "*" application identifier, to allow them to set arbitrary application identifiers, and install those on provisioned devices. If you wish to use Apple's push notifications, however, you will not be able to use wildcards—it's Apple's policy that the provisioning profile has to match the application identifier exactly, in this case.
- Ensure you have all of your development devices on every provisioning profile that you use regularly. This will avoid confusing errors when you want to test a new project on an older device that you have.
- As far as possible, ensure that the Mac you use for development is the same one you'll use to develop your release build and submit your application. For single developers, this is not often an issue, but if you're working on a team (and using your company's development certificate), this is liable to catch you out at an inopportune time, and your application will not be submitted to Apple's App Store.
- One more point to note—the iOS Provisioning Portal currently allows for a maximum of 100 devices to be registered per account. This may sound like a lot for a single developer, but if you have a larger company or set of clients, it may get a bit tight—you'll want to look into Apple's **Ad Hoc** distribution methods. However, this won't be a concern during your initial development.

Thankfully, while using PhoneGap we can work on much of the functionality of our applications using desktop web browsers, which means we do not need to deploy to devices until late in the process.

Have a go hero–iOS basics

Although we won't be using the standard iOS development tools—namely, coding with **Objective-C** and designing with **Interface Builder**—it's good to get a firm grasp of the nuts and bolts of iOS development.

You might want to see if you can change the name of `FirstApp`; perhaps rename it to `App The First`. Are you able to change the icon? Can you create an application—`The First Lite project`, and install it on your simulator and device alongside `App The First`?

B
Pop Quiz Answers

Chapter 1

PhoneGap iPhone Basics Answers

Where are your PhoneGap assets (HTML, JavaScript, and CSS) located in an Xcode project?

Answer#3— In the `www` folder. `www` is the standard folder for all PhoneGap platforms.

How do you rename a PhoneGap iOS application?

Answer#3— Edit the `Application-Info.plist` file, like any other native iOS app.

What function is called by the alert('Hello PhoneGap!') code?

Answer#1— The standard `alert` function in the iOS WebView. PhoneGap uses native JavaScript calls wherever feasible.

Chapter 2

Initial Design Answers

Let's say we wanted to add a second page to our application, with the same CSS and JavaScript code. What would be the best approach?

Answer#2 is the most manageable and maintainable.

Answer #1 introduces more work, while #3 is fantasy.

You Are The Best *has proved so popular that the users are demanding a tablet application release to go with their smartphone application. How should we edit the CSS?*

There's no absolute right answer here— For an application this simple, a combination of #1 and #2 would be okay. For a larger project, something like #3 is essential, probably using CSS3 media queries.

The other demand we've received from users is for a second button, that says **Tell my cat she's great too**. *This button should present a different message to the user. How should we best implement this code?*

All three examples would work answer #1 is the cleanest way to set things up, and the most idiomatic JavaScript.

Chapter 3

Templating with Mustache Answers

What advantages does Mustache have over simple string interpolation?

I hate to do this to you but, even though the appearances are a little contradictory, the correct answer is all of the above. In terms of technical facility, there's nothing Mustache does that any determined programmer couldn't do, but the benefit of having it already done, and done right, is the major benefit for web developers.

When is the best time to render a template for display?

Well answer #2 is out of the question–detecting when the user is idle is an especially quixotic goal, made all the complicated by resource-constrained mobile devices. There are certain circumstances when answer #1 is suitable–for example, with a modal view (such as a Lightbox) that could appear at any stage in the application. In most cases, answer #3 is correct, but if the view rendering takes a long time, make sure you give some response to the user right away.

Chapter 4

Feature Detection vs UA Sniffing Answers

For each of these examples, would you be able to use feature detection or would you have to fall back to user agent sniffing?

- Checking if the device has a hardware back button.
- User agent sniffing: No browser currently exposes the amount of hardware buttons in any meaningful way. Thankfully, this is a good case for UA sniffing: the quantity of hardware buttons is not something that will change with an OS upgrade, only with new hardware or platforms.
- Finding support for a WebSQL SQLite database.
- Feature detection: if WebSQL is supported, a top-level openDatabase function is attached to the window object.
- Finding support for multitouch events: This is a tricky one: we can simulate a touch event and check for a `touches` array in the resulting event object. If multitouch is supported the `touches` array contains a set of points corresponding to each touch. However, on Android, at the time of writing, the `touches` array is only ever populated with a single touch, though multitouch is supported for pinching and zooming web pages as a whole. As with hardware accelerated transforms, this multitouch support requires either inaccurate feature detection or brittle user agent sniffing.

  Checking the total memory available to your application: Neither—this data is not exposed by the user agent, and is not available to the DOM. If this is a pressing concern for your application, you probably have bigger problems to worry about for the moment.

When performing feature detection, when should we execute the detection code?

Answer #3 is correct, with the caveat that the result should be cached for future access, not recalculated every time the feature is accessed. As mobile developers, we should emphasize quick start-up times, and defer as much as possible until later in the application's life-cycle (this is one argument against using a library like Modernizr, instead of performing your own ad hoc feature detection while the application is running).

What kind of code branching should you do with user agent sniffing?

Go as broad as possible—answer #1 is the best bet. Since user agent sniffing is such a brittle technique, you don't want to use it for anything that's especially variable, such as different build versions of the browser engine, or even different devices on a single platform.

The exception to this would be platforms that have drastic differences between devices—for example, the PhoneGap BlackBerry widgets project supports both BlackBerry OS 5.0, with RIM's proprietary browser in place, and BlackBerry OS 6.0, which uses a far more capable WebKit browser. However, the differences between these versions of the OS are so pronounced that feature detection will take you most of the way.

Chapter 5

XUI Answers

Instead of removing food items from our page, let's say we wanted to hide them, with the option of bringing them back later. What would be the best way of doing this (assume we have selected the correct element, and hidden is a CSS class with display:none as a property)?

Answer #2. `hide()` is not a function in XUI (although it's easy enough to add as a convenience function, as it exists in jQuery). Answers #2 and #3 are functionally identical, but answer #2 has the benefit of being able to easily retrieve the hidden items (`by x$('.hidden))`.

Is XUI compatible with jQuery plugins?

Answer #2 is correct. It is possible that some jQuery plugins use a subset of jQuery's functionality that coincide with XUI, but there are no guarantees that the semantics of functions with the same name would be compatible. Many would be compatible with a few tweaks–if the plugin has a test suite, it's worth having a look.

What's the primary advantage XUI has over other JavaScript (DOM) libraries?

The correct answer is #1. Answer #2 is generally true—XUI tends to use native DOM methods under the hood, which are faster than other libraries' custom code (for CSS selectors, say— but it's not to the extent that one would drop any other library). If jQuery compatible syntax is the priority, then jQuery is your best bet.

Media Elements Pop Quiz Answers

Which of the following is NOT possible using the HTML5 video API, with a thirty-second long video clip?

The correct answer is #2. Answer #1 can be achieved through a judicious use of the `currentTime` attribute and `setTimeout`. #3 works just by using `currentTime` and play.

What is the difference between the `canplay` *and* `canplaythrough` *events?*

Answer #3

If a video is set to `autobuffer`*, when does it start to download?*

Answer #2. Answer #1 occurs when `autobuffer` is not set (it buffers manually). Answer #3 is a special case of #2—if the video tag is in the DOM when the page loads, the video will start buffering. If, as in our example, the video tag is not in the DOM, then it will not buffer until it is.

Chapter 6

Scrolling Answers

Why doesn't `position: fixed` *work on many mobile browsers?*

Answer #2 is correct—see the viewport section above. With scrolling and with user scaling of the page, it would be difficult for `position: fixed` to make sense at it does on desktop browsers.

What does iScroll use CSS3 transforms for?

Answer #3. The other two answers are important parts of iScroll's functionality, but do not need CSS3 to behave correctly.

Can I forego using a scrolling library, and just use the web view's scrolling facilities?

Answer #3–that is, yes and no. It will depend on the demands of your particular application and its users. In many cases, using a scrolling library is a necessary evil, but it's by no means mandatory.

Chapter 7

Geolocation Answers

When is the user prompted to allow geolocation access?

Answer #2—when the user calls `getCurrentPosition`. Because of this, you should ensure that the user is aware they will be prompted for location data, so the prompt does not disturb their experience of your application.

What happens when the user refuses to allow access to their location data?

Answer #1—when the error `callback` is called. Because of this, it is especially important to pass an error `callback` to `getCurrentPosition`, so there's no unexpected behavior regardless of the user's action.

How is geolocation data calculated?

Answer #3–as an application developer, you just do not know. If the quality of the reading is a key component of your application's functionality, always ensure that you check the accuracy property of the location object.

Orientation and Media Queries Answers

What is the main benefit of CSS media queries?

Answer #2— Answer #1 is possible, but there have been mixed reports about whether browsers respect the media attribute when choosing which stylesheets to download. In the case of device orientation, the user would need both sets of styles at any rate.

Why is `orientation` preferred *to* `min-device-width`*, in our example?*

Answer #3— Using `min-device-width` is a lot like user agent sniffing in this regard, and has many of the same drawbacks.

Is it necessary to support all orientations on a device?

No. As we have seen, it is the default behavior on iPhone PhoneGap applications to only support portrait orientation. You can configure Android applications similarly.

Chapter 8

navigator.camera.getPicture Answers

Why can we not mock `navigator.camera.getPicture` *on a desktop browser?*

Answer #3—there is no access to device cameras. This will change in the future ; there is a **Media Capture** specification being worked on—but that interface will be more like the `<input type="file">` tag than the PhoneGap camera API.

Which is the most robust data source for retrieving images?

Answer #1—this will bring up some interface on any PhoneGap platform that supports the camera API.

How is the user interface for the image picker defined?

Answer #3 is correct—this is especially thorny on Android. Different device manufacturers or carriers can override the default image pickers, and users are able to as well. Test on devices!

Destination Types Answers

Select which `destinationType` *is best suited for each of the following applications:*

1. *A photo-sharing application, which sends images to the user's friends—*`DATA_URL`*–we can't post* `FILE_URI`*s to a remote server.*
2. *A high-resolution image viewer, where users can zoom into the high-megapixel pictures taken with their camera—*`FILE_URI`*—*`DATA_URL` *can be unstable with large image data.*
3. *An image manipulation application, allowing images to be stretched and squeezed with a multitouch interface—This one is debatable – you might want a higher image quality and favour* `FILE_URI`*, while* `DATA_URL` *can be possibly easier to manipulate. In most* cases, `DATA_URL` *would be better.*

Why do data-uri image sources require the `data:image/png;base64`*, prefix?*

Answers #1 and #2 are both correct— We need to ensure the data is parsed as image data, not a file path, and we also need to ensure that it is parsed as a `.png` image, not any other kind.

Why are file paths not a good choice for persistent data?

Answer#2—If the data is important to our application, we want to have control over it. Answer #1 is true but irrelevant; the same images won't be on different devices anyway. Answer #3 is wrong because PhoneGap won't return a path that your application cannot access.

Chapter 9

Contacts Answers

Which fields need to be set before `contact.save()` *can be called?*

Answer #3, but that wouldn't be much use. A good rule of thumb is to follow answer #2, to ensure that on every platform where the contact is created, the user can identify it (to edit/augment the contact, or to delete it).

Why is it a good idea to set a filter when calling `contacts.find()`?

Answer #1 is the best reason, since a crash is the worst thing that can happen to your application It's also true that, in most cases, returning all of the user's contacts is a bit of a pointless exercise.

Why are `phoneNumbers` and `emails` set as arrays, rather than string fields?

Answer #2. But partial credit if you said #3.

Chapter 10

Using PhoneGap Plugins Answers

What best describes the process of adding a PhoneGap plugin to your application?

Sadly, answer #2 is currently the most reliable option. There isn't a strongly enforced standard for PhoneGap plugins right now (though that may have changed somewhat by the time you read this). Answer #3 is the goal once the plugins standardize a bit; #1 is technically possible, but no plugins adopt that strategy right now.

Why does the ChildBrowser plugin behave differently on iOS, compared to Android?

The most accurate answer is #1—While on Android it's acceptable for an application to use an `Intent` and open a new view, on iPhone users expect that such links will open in the context of the current application. #2 is also true, but both ports have the same aim (show an external web page without disrupting the user's in-app experience).

PhoneGap plugins require native code to be written independently for each platform. Why is this?

Answer #2. Answer #1 is accurate, in the case of Android and iOS, but it fails to account for platforms that have the same programming language, but different underlying APIs (for example, Android and BlackBerry). You could theoretically build a cross-compatible plugin for these platforms (a pure math library, perhaps), but most PhoneGap plugins are interested in the native APIs, which will be platform-specific.

Writing PhoneGap Plugins Answers

Why should you start with a single JavaScript API for a plugin, and write your native implementations after that?

Answer #1, code reuse. In addition, having a settled JavaScript API gives you less to worry about when writing native code—you simply need to get the native code interacting to the same high-level interface.

What is the main difference between plugins on Android and BlackBerry WebWorks?

The underlying platform APIs, as demonstrated in the Battery plugin example. There may well be differences between the Java VM and the Dalvik VM, but they have not hugely affected PhoneGap developers in any noticeable way.

Can you write native code for your application without using plugins?

Answer #2, absolutely. If you find the ceremony of the `Plugin` and `PluginResult` classes too much to bear, it's straightforward to borrow the PhoneGap techniques for your own ends. The plugin interface is just convenient and helps you to do this in a stable and easy-to-follow fashion.

Chapter 11

A Simple Web Service Answers

Why does the service offer the news stories in JSON format, as well as HTML?

All of the answers are correct—as a data transfer format, JSON is more easily processed than HTML, is generally lighter in terms of bandwidth, and does not contain extraneous presentational information. It's also trivial to implement, at least in our example, so there's no good reason in this case to send rendered HTML over the wire.

Is our **News Stories** *web service ready for a production deployment?*

Well, probably not, due to the reasons stated in answer #2. For any application you would deploy to the public internet, you would minimally need to sanitize any user input that comes in before displaying it back to the user. In our case, it's not a real concern: we don't have a database that's vulnerable to SQL injection, nor do we have any user data that is vulnerable to a cross-site scripting exploit. But the service could conceivably (bear with me here) grow to the extent that it would have sensitive content; you would want to take these issues into consideration as early as possible.

What is the best approach for loading our news stories from our service?

This one is interesting because the best answer for PhoneGap applications is #1— Loading through an `XmlHttpRequest`—is not applicable for mobile websites in general, which cannot access external resources in this way, due to the same-origin policy on websites. If we were writing a service for general consumption, it would be worth the extra effort to implement JSON support.

Index

Symbols

A

B

C

D

L

M

N

O

P

Q

R

S

T

V

W

X

Y

[PACKT] open source
community experience distilled
PUBLISHING

Thank you for buying

PhoneGap Beginner's Guide

About Packt Publishing

Packt, pronounced 'packed', published its first book "*Mastering phpMyAdmin for Effective MySQL Management*" in April 2004 and subsequently continued to specialize in publishing highly focused books on specific technologies and solutions.

Our books and publications share the experiences of your fellow IT professionals in adapting and customizing today's systems, applications, and frameworks. Our solution based books give you the knowledge and power to customize the software and technologies you're using to get the job done. Packt books are more specific and less general than the IT books you have seen in the past. Our unique business model allows us to bring you more focused information, giving you more of what you need to know, and less of what you don't.

Packt is a modern, yet unique publishing company, which focuses on producing quality, cutting-edge books for communities of developers, administrators, and newbies alike. For more information, please visit our website: `www.packtpub.com`.

About Packt Open Source

In 2010, Packt launched two new brands, Packt Open Source and Packt Enterprise, in order to continue its focus on specialization. This book is part of the Packt Open Source brand, home to books published on software built around Open Source licences, and offering information to anybody from advanced developers to budding web designers. The Open Source brand also runs Packt's Open Source Royalty Scheme, by which Packt gives a royalty to each Open Source project about whose software a book is sold.

Writing for Packt

We welcome all inquiries from people who are interested in authoring. Book proposals should be sent to author@packtpub.com. If your book idea is still at an early stage and you would like to discuss it first before writing a formal book proposal, contact us; one of our commissioning editors will get in touch with you.

We're not just looking for published authors; if you have strong technical skills but no writing experience, our experienced editors can help you develop a writing career, or simply get some additional reward for your expertise.

[PACKT] open source
PUBLISHING
community experience distilled

Cacti 0.8 Beginner's Guide

ISBN: 978-1-849513-92-0 Paperback: 348 pages

Learn Cacti and design a robust Network Operations Center

1. A complete Cacti book that focuses on the basics as well as the advanced concepts you need to know for implementing a Network Operations Center

2. A step-by-step Beginner's Guide with detailed instructions on how to create and implement custom plugins

3. Real-world examples, which you can explore and make modifications to as you go

4. Written by Thomas Urban – creator of the "Network Management Inventory Database" plugins for Cacti

---

MeeGo 1.0 Mobile Application Development Cookbook

ISBN: 978-1-849690-32-4 Paperback: 300 pages

Simple and effective recipes for professional MeeGo mobile applications supporting calls, SMS, UI, display, GPS, multimedia, and much more

1. A step-by-step guide to creating feature-rich, powerful Qt mobile applications in Python rapidly

2. Quick recipes for building professional Smartphone applications for UI, display, GPS, multimedia, and games

3. Plenty of code examples to help you develop your own applications

4. The only book to cover common MeeGo mobile application development problems and smart solutions

Please check **www.PacktPub.com** for information on our titles